Tackling Turbulent Flows in Engineering

Anupam Dewan

Tackling Turbulent Flows in Engineering

 Springer

Prof. Dr. Anupam Dewan
Department of Applied Mechanics
Indian Institute of Technology Delhi
110016 Hauz Khas, New Delhi
India
e-mail: adewan@am.iitd.ac.in

ISBN 978-3-642-42863-0

ISBN 978-3-642-14767-8 (eBook)

DOI 10.1007/978-3-642-14767-8

Springer Heidelberg Dordrecht London New York

Cover design: WMXDesign GmbH

Printed on acid-free paper

Springer is part of Springer Science+Business Media (www.springer.com)

Preface

As the name implies, this book presents different ways to deal with turbulent flows that are encountered in a wide spectrum of engineering applications. The objective is to provide guidelines to engineers of several disciplines on tackling turbulent flows with the use of minimum possible number of mathematical equations. The present book offers several novelties: it presents (a) engineering aspects of turbulent flows; (b) ways to handle turbulent convective heat transfer applications including both forced and free convections; (c) examples of some important practical situations in which turbulent flows are encountered.

Chapter 1 presents the basics of fluid mechanics and convective heat transfer. This chapter is expected to lay the foundation for understanding the material presented in the remaining chapters of the book. Chapter 2 presents an introduction to turbulent flows. It also describes properties of turbulent flows. Chapter 3 presents characteristics of some important turbulent flows, which include, boundary-layer flows, different types of free-shear flows and flow through a circular pipe. Chapter 4 presents a widely used approach to treat turbulent flows, that is, the use of time-averaged governing equations. It also shows that this approach results in an additional complexity termed as the closure problem. Chapter 5 presents different turbulent models based on the widely used Boussinesq approximation that can be adopted to close the system of time-averaged governing equations. Chapter 6 presents the details of the widely used standard k-ε model and some other two equation models. Chapter 7 presents the physically most rigorous method to handle time-averaged governing equations, that is, Reynolds-stress and scalar flux transport model. Chapter 8 presents another entirely different approach for turbulent flows, which involves a treatment of three-dimensional, instantaneous flow field using direct numerical simulation and large eddy simulation. Chapter 8 also presents different subgrid scale models for large eddy simulation and it also shows why it is difficult to use these two techniques as a design tool for engineering applications. Chapter 9 presents nine examples of turbulent flows from the literature covering a wide range, which include ventilation in buildings, stirred vessels used in chemical industries, heat exchangers, tundish used in steel industry, particle deposition in a human throat. Chapter 10 presents conclusions and issues related to computational simulation of engineering turbulent flows.

The book is an outcome of my experience with the interesting subject of fluid mechanics in general and turbulence modeling in particular during the past two decades. I started appreciating the complexity of turbulence flows when I joined for Master of Technology degree at Department of Mechanical Engineering, Indian Institute of Science Bangalore and was fortunate of have Professor Jaywant H. Arakeri and Professor J. Srinivasan as my project supervisors. My appreciation continued to get strengthened while I did my Ph.D. at the same institution with the same supervisors. I thank my these two former supervisors and all my teachers at Indian Institute of Science, Bangalore, especially Professor Vijay H. Arakeri, Department of Mechanical Engineering for their excellent teaching.

A major part of this book was written from March 2008 to May 2009 while I was a Professor of Mechanical Engineering at Indian Institute of Technology Guwahati. During this time, I had the opportunity of teaching a related course entitled "Numerical Simulation and Modelling of Turbulent Flows" to B.Tech., M.Tech. and Ph.D. students, mostly from the Department of Mechanical Engineering. I thank all my students who have done this course at Indian Institute of Technology Guwahati. Their active participation in the class discussions helped me effectively understand this subject and improve the presentation of some difficult topics of this book. I thank the administration of Indian Institute of Technology Guwahati and all my former colleagues at Department of Mechanical Engineering for all their help during the book writing. I especially thank my friend and former colleague Professor Anoop Kumar Dass at Indian Institute of Technology Guwahati with whom I had numerous stimulating discussions on this subject during my employment with Indian Institute of Technology Guwahati for nearly 12 years. I thank Mr. Nandan K. Das, Indian Institute of Technology Guwahati for drawing all figures of this book.

I gave finishing touches to this book during my subsequent employment as Professor of Applied Mechanics, Indian Institute of Technology Delhi. I thank the administration of Indian Institute of Technology Delhi for creating a conducive atmosphere to write this book. I especially thank my current Ph.D. student Mr. Rabijit Dutta, who very patiently read the entire manuscript, found several typographical errors and pointed out the need for improvements in presentations at several places in the book.

I thank Springer-Verlag GmbH for readily agreeing to my proposal for this book. I also thank Dr. Christoph Baumann, Editor, Springer-Verlag GmbH, for patiently agreeing to my requests for extension in the submission date and also for promptly responding to my numerous queries.

I thank my wife Meenakshi for her continuous support and encouragement during the entire writing process. This book would not have been possible without her. I also thank my two lovely daughters Rimjhim and Sanya who very patiently tolerated long periods of my absence during weekends and holidays over the last two years. I also thank my parents Dr. Manohar Lal Dewan and Mrs. Sudershan

Dewan for their continuous encouragement and sacrifice towards my education right from my primary school days. I also thank my brother Dr. Deepak Dewan and sisters Dr. Sangeeta Khanna and Dr. Mohini Puri for their support.

Contents

Chapter 1
Basics of Fluid Mechanics and Convective Heat Transfer

Abstract In this chapter we present basics of fluid mechanics which include concept of fluid, important fluid properties, different methods of flow visualization, difference between Eulerian and lagrangian approaches, between integral and differential treatments. We present differential equations for the conservation of mass, momentum and energy and the corresponding boundary conditions. We also introduce the reader to convective heat transfer. The chapter concludes with two examples of flow normal to an infinite circular cylinder and convective heat transfer from a heated tube surface to a cold fluid flowing axially within the tube.

1.1 Fluid Properties

A fluid deforms continuously when a tangential stress is applied irrespective of the amount of stress. A solid can be distinguished from liquid based on its response to an applied stress. A solid can resist a shear stress by a static deformation, but a fluid cannot resist a shear stress. Both liquids and gases are treated as fluid. A fluid can be treated as continuum in most situations, except at extremely low pressure where the molecular spacing and mean free paths are comparable to the physical dimensions of the flow. Fluid mechanics is a broad subject. In the subsequent sections we will consider important physical properties of fluids that have an influence on their fluid and thermal behaviours.

1.1.1 Viscosity

Viscosity is probably one of the most important fluid properties. It is a measure of the ability of fluid to resist flow. Viscosity results in adjacent moving fluid layers having relative velocities and this is termed as the shearing of layers. Viscosity is a thermodynamic property and therefore changes with a change in thermodynamic state of fluid (for example, with a change in pressure and temperature). However,

A. Dewan, *Tackling Turbulent Flows in Engineering*,
DOI: 10.1007/978-3-642-14767-8_1, © Springer-Verlag Berlin Heidelberg 2011

its variation with pressure can be practically neglected. Temperature has a rather strong effect on viscosity. Viscosity increases with temperature for gases and decreases for liquids. All real fluids have a finite value of the viscosity. A fluid having the zero viscosity is an assumption and such fluids are termed as the ideal fluids. We will see in Sect. 1.3 that the assumption of zero viscosity leads to a considerable simplification in the treatment of fluid behaviour.

Reynolds number (Re) is an important non-dimensional parameter for viscous flows and is defined as the ratio of the inertia forces to the viscous forces.

$$Re = \frac{\rho V L}{\mu} \tag{1.1}$$

where ρ denotes the density, V a characteristic velocity, L a characteristic length scale and μ the coefficient of viscosity. Creeping flows are characterized by extremely low values of Reynolds number. These flows are dominated by the viscous effects. A smooth laminar flow is characterized by moderate values of Reynolds number and disorderly fluctuating turbulent flows are characterized by high value of Reynolds number. The present book is devoted to modeling and simulation of turbulent fluid flows and convective heat transfer.

1.1.2 Density

Density denoted by symbol ρ is defined as the mass per unit volume. It is an intensive (i.e., per unit mass) thermodynamic property. Density of liquids can be assumed to be constant and thus liquids can ordinarily be assumed to be incompressible. However, density varies considerably with pressure in a gas and one needs to consider the value of Mach number before assuming a gas to be incompressible. A variable density leads to complexity by addition of another variable, namely, density (ρ) to the flow behaviour. In the present book we will assume fluids to be incompressible while treating turbulence.

1.1.3 Thermal Conductivity

Thermal conductivity is a diffusive property which results in the smoothening of temperature gradients or in other words diffusion of temperature. According to the Fourier's law of heat conduction the relation between the heat flux and temperature gradient is expressed as

$$\vec{q} = -k\vec{\nabla}T \tag{1.2}$$

Here $\vec{q}$ denotes the heat flux and k the thermal conductivity. The negative sign takes care of the direction of heat transfer which is in the direction of decreasing temperature. Thermal conductivity (k) is a thermodynamic property. Heat flux $\vec{q}$ is a vector quantity and can also be written in the expanded form in different directions according a particular coordinate system being used.

1.1.4 Surface Tension

Surface tension arises when a liquid is in contact with a liquid, gas, vapour or solid. It is represented as force per unit length. Standard data is available for surface tension for different pairs of liquids in contact with different clean solid surfaces. These values change significantly if the surface gets contaminated. Sometimes a curved liquid surface is formed in contact with a solid, for example, for clean mercury–air–glass interface, the contact angle is 130°. Surface tension needs to be taken into account when considering force balance and the contact angle also needs to be considered in the analysis to obtain components of surface tension in different co-ordinate directions.

1.1.5 Speed of Sound and Mach Number

The speed of sound of a fluid a is the rate of propagation of small pressure pulses through the fluid and is defined as

$$a^2 = \left(\frac{\partial p}{\partial \rho}\right)_s \qquad (1.3)$$

Here subscript s denotes an isentropic (i.e., constant entropy) process. For an ideal gas the expression of the speed of sound simplifies to

$$a = \sqrt{\gamma RT} \qquad (1.4)$$

Where γ denotes the ratio of specific heats, R the gas constant and T the temperature in K.

Mach number Ma is defined as the ratio of local fluid velocity V and the speed of sound of fluid a ($Ma = V/a$). Compressible flows are important in aerodynamics and turbo-machinery. Typically a flow can be assumed to be incompressible for $Ma < 0.3$ and therefore for $Ma > 0.3$ a variation in the density needs to be considered.

1.1.6 Newtonian and Non-Newtonian Fluids

Newtonian fluids are the ones in which the stress is linearly proportional to rate of strain.

$$\tau \propto \frac{\delta\theta}{\delta t} \qquad\qquad (1.5)$$

Here τ denotes the stress and θ the strain angle which can be written in terms of the flow field. The constant of proportionality is the coefficient of viscosity. Fluids that do not follow a linear relation between the stress rate and the corresponding rate of strain are termed as the non-newtonian fluids. Some examples of non-newtonian fluids are dilatant, pseudoplastic and plastic. Newtonian fluids only will be treated in the present book.

1.2 Treatments and Visualization of Fluid

Sometimes one may gather important information about flow behavior in a particular situation without actually measuring or computing any quantity. In this section we will treat different ways that can be used to visualize a fluid flow.

1.2.1 Eulerian and Lagrangian Approaches

Eulerian approach is also called the field approach, wherein one is interested in the flow field within a fixed region in space and not the changes in flow properties that a particular fluid particle moving through a flow experiences. This approach is consistent with our expectations from a typical treatment of fluid flows. For example, we would like to know how much drag needs to be overcome while designing an airfoil or how much pumping power is required to enable a flow of a particular liquid take place through a pipe of a particular length and diameter. In other words, we are interested in a particular region in space. In the lagrangian approach, we are interested in the behaviour of an individual fluid particle as it moves through a flow field. The later approach is suitable in solid mechanics and is generally not used in fluid mechanics.

1.2.2 Streamline, Streakline and Pathline

Different methods can be adopted to visualize the flows. A streamline is a line tangent to the velocity vector at a given instant (Fig. 1.1). The flow can take place only along a streamline and it is clear that there can be no flow across a streamline. The volume flow rate between any two streamlines can be written as the difference between the constant values of stream functions (defined in the next paragraph) defining the two streamlines. A pathline is the actual path taken by a particular fluid particle over a period of time. A streakline is a locus of all particles that have earlier passed through a prescribed point. A timeline is a set of fluid particles that form a line at a given instant. Streamline and timeline are instantaneous lines

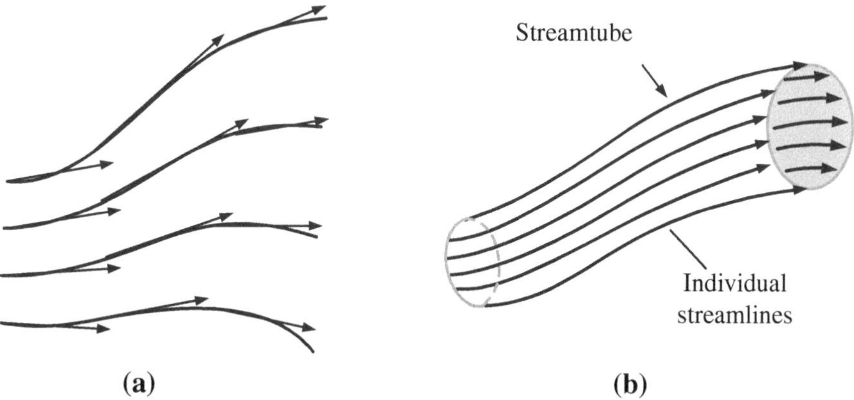

Fig. 1.1 **a** A schematic of streamlines in a two-dimensional flow and **b** streamtube

whereas streakline and path-line are generated over a time period. Streamlines, pathlines and streaklines are identical in a steady flow, a flow field whose behaviour does not change with time. Streamlines are widely used for the visualization in fluid mechanics (Fig. 1.1a). A streamtube is a set of streamlines forming a tube like structure (Fig. 1.1b). A flow cannot cross the boundaries of a streamtube and therefore it is clear that a streamtube walls are not solid but fluid surfaces.

A stream function ψ can be defined for a particular streamline. However, it can be defined for steady, two-dimensional flows only in such a way that

$$u = \frac{\partial \psi}{\partial y} \text{ and } v = -\frac{\partial \psi}{\partial x} \tag{1.6}$$

where u and v denote velocity components in x- and y-directions, respectively. Thus the continuity equation for the two-dimensional steady flow is identically satisfied by the above-mentioned substitution. In other words, stream function satisfies the continuity equation for two-dimensional, incompressible flows.

1.2.3 Integral and Differential Treatments

In the integral approach a finite region in space is considered and not discrete points in space. The region is termed as control volume and is identified by a control surface. This approach is also termed as the control volume approach and one considers gross effects within the control volume, for example, how much heat is crossing the surface of a heat exchanger or how much work is obtained from a turbine. In Chap. 3 using the examples of turbulent fully developed flow through a circular pipe and turbulent flow past a flat plate we will see that in some situations we can obtain extremely useful information from the integral analysis. In the differential approach, fluid behaviour is considered at each point in the flow field. The second approach will be used in the present book and the relevant details are presented in the subsequent sections.

A general flow field is a function of both space co-ordinates (x, y and z) and time (t), i.e., is three-dimensional and unsteady and therefore quite difficult to analyze. Therefore in many situations it is good to reduce the dependence of the flow field on the independent variables. A flow field can be termed as one-, two- or three-dimensional depending on the number of space co-ordinates required to completely specify the flow field. For example, a fully developed laminar flow through a circular tube can be assumed to be one-dimensional because it varies only in the radial direction. However, a developing but axisymmetric flow through a circular tube can be assumed to be two-dimensional because it varies in the axial as well as the radial directions.

1.3 Vorticity and Irrotational Flow

A fluid particle moving in a general three-dimensional flow field may rotate about all three co-ordinate axes. Fluid rotation is a vector quantity and is defined as

$$\omega = i\omega_x + j\omega_y + k\omega_z \tag{1.7}$$

where ω_x, ω_y and ω_z denote the rotation about x-, y- and z-axes, respectively.

Rotation (ω) is defined as half of the curl of the velocity vector $\omega = Curl\ \vec{V}$ and can be written as

$$\omega = \frac{1}{2}\begin{vmatrix} i & j & k \\ \frac{\partial}{\partial x} & \frac{\partial}{\partial y} & \frac{\partial}{\partial z} \\ u & v & w \end{vmatrix} \tag{1.8}$$

Another form of the rotation is vorticity (ξ) and is given as $\xi = 2\omega$. A flow with zero angular velocity ($\omega = 0$) is termed as the irrotational flow. Velocity potential (ϕ) can be defined for irrotational flows. It is a three-dimensional function and the velocity components can be readily calculated from the velocity potential. It can be shown that the velocity potential and stream function are normal to each other except at the stagnation point.

All three components of fluid rotation are nonzero in a general three-dimensional flow. However, even in a two-dimensional flow in the x–y plane, the fluid rotation about the z-axis given by $\omega_z = \frac{1}{2}\left(\frac{\partial v}{\partial x} - \frac{\partial u}{\partial y}\right)$ may be non-zero and therefore the flow may be rotational. Viscosity is a major cause of the generation of the vorticity (or rotation). However, in many viscous flows, the vorticity may be negligible and the flow may be assumed to be irrotational. For example, Prandtl proposed that it is adequate to consider thin viscous region close to a solid surface, termed as the boundary layer, to account for the viscous effects. The remaining rather large region can be considered to be inviscid or irrotational and thus simplified governing equations can be adopted in such thin regions. We will consider these equations in the subsequent sections.

Substitution for u and v in terms of the stream function and velocity potential in the continuity equation for two-dimensional incompressible flow results in the following equations

$$\frac{\partial^2 \phi}{\partial x^2} + \frac{\partial^2 \phi}{\partial y^2} = 0 \tag{1.9}$$

$$\frac{\partial^2 \psi}{\partial x^2} + \frac{\partial^2 \psi}{\partial y^2} = 0 \tag{1.10}$$

These equations are termed as the Laplace's equation which arises in many situations in science and engineering. A treatment of these equations is simple because of their linear nature.

1.4 Force, Strain and Stress

Surface forces act on the surface of a body through direct contact with a surface. Forces acting over the volume of fluid are termed as the body forces. Gravitational force is an example of a body force.

Strain in fluid results in the stresses, both normal and shear. The stress at a point in a flow field is specified by nine components. The state of stress can be described completely by describing three stresses acting on three mutually perpendicular planes each passing through the point of interest (total nine components).

Consider two layers of fluid moving parallel to each other at velocities u and $u + \delta u$ (Fig. 1.2). The shear strain angle will grow continuously for the duration of application of shear stress τ. For the Newtonian fluids it can be shown that

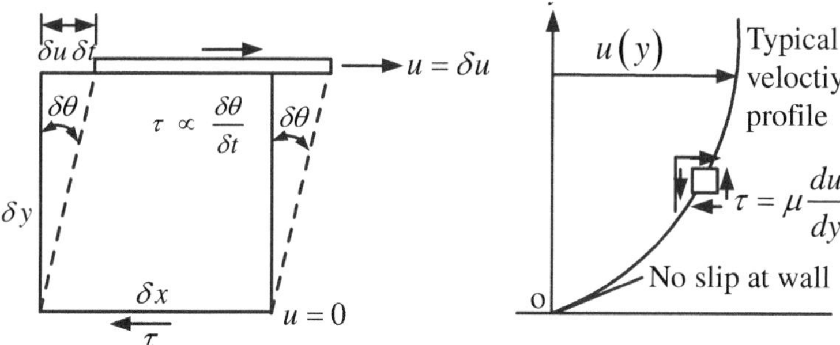

Fig. 1.2 Deformation in a fluid due to shear

$$\tau = \mu\frac{\mathrm{d}\theta}{\mathrm{d}t} = \mu\frac{\mathrm{d}u}{\mathrm{d}y} \qquad (1.11)$$

Equation 1.11 is applicable to the specific flow shown in Fig. 1.2. It can be shown the general expressions for the shear and normal stresses in the component form in a Cartesian co-ordinate system for an incompressible fluid can be written as (White 2003).

Normal stresses:

$$\tau_{xx} = 2\mu\frac{\partial u}{\partial x}, \quad \tau_{yy} = 2\mu\frac{\partial v}{\partial y}, \quad \tau_{zz} = 2\mu\frac{\partial w}{\partial z} \qquad (1.12)$$

Shear stresses:

$$\tau_{xy} = \tau_{yx} = \mu\left(\frac{\partial u}{\partial y} + \frac{\partial v}{\partial x}\right), \quad \tau_{xz} = \tau_{zx} = \mu\left(\frac{\partial w}{\partial x} + \frac{\partial u}{\partial z}\right),$$
$$\tau_{yz} = \tau_{zy} = \mu\left(\frac{\partial v}{\partial z} + \frac{\partial w}{\partial y}\right) \qquad (1.13)$$

The corresponding expressions for a compressible fluid are somewhat more complex compared to those in Eq. 1.13. Stress at a point is denoted by two indices. The first one represents the direction of the normal to the plane and second index represents the direction of stress. A stress component is treated as positive when the direction of the stress and the plane on which it acts are both positive or both negative. A stress component is considered negative when the two above-mentioned directions (of plane and stress) are opposite to each other (Fig. 1.3). Pressure is a compressive normal stress at point in a fluid and therefore needs to be accounted for along with normal stresses when considering the balance of forces and therefore the net stresses can be written in the matrix form as

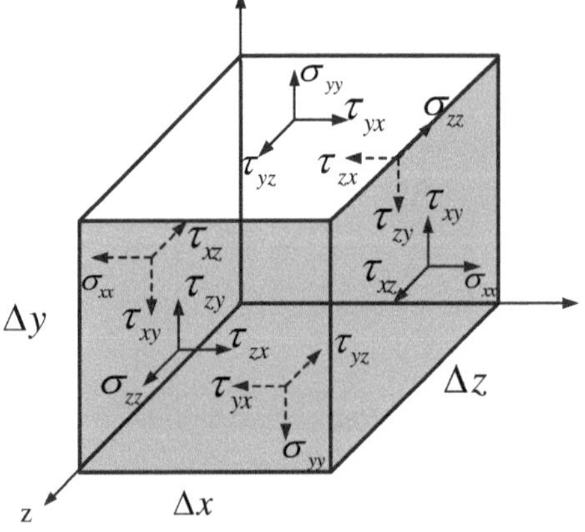

Fig. 1.3 Schematic of notation used for describing the stresses in a fluid

$$\sigma_{ij} = \begin{vmatrix} -p + \tau_{xx} & \tau_{yx} & \tau_{zx} \\ \tau_{xy} & -p + \tau_{yy} & \tau_{zy} \\ \tau_{xz} & \tau_{yx} & -p + \tau_{zz} \end{vmatrix}$$

Pressure gradient is a surface force that acts on the sides of a fluid element.

1.5 Fluid Acceleration

We need to compute the fluid acceleration to apply the Newton's law of motion to an infinitesimal control volume. It can be obtained from the flow field $\vec{V}(x, y, z, t)$ by applying the chain rule. Fluid acceleration is a vector and can be defined in the component form in the Cartesian coordinate system as

$$a_x = \frac{\partial u}{\partial t} + u\frac{\partial u}{\partial x} + v\frac{\partial u}{\partial y} + w\frac{\partial u}{\partial z} \tag{1.14}$$

$$a_y = \frac{\partial v}{\partial t} + u\frac{\partial v}{\partial x} + v\frac{\partial v}{\partial y} + w\frac{\partial v}{\partial z} \tag{1.15}$$

$$a_z = \frac{\partial w}{\partial t} + u\frac{\partial w}{\partial x} + v\frac{\partial w}{\partial y} + w\frac{\partial w}{\partial z} \tag{1.16}$$

where u, v and w denote velocity components in the x, y and z directions, respectively. Each component of fluid acceleration consists of two terms: local acceleration (first term on the RHS of Eqs. 1.14–1.16) and the convective acceleration (last three terms on the RHS of Eqs. 1.14–1.16). Local acceleration is zero in a steady flow, which is the one in which the flow properties do not change with time. The second term arises due to spatial variations in the velocity field. For example, a flow through a nozzle results in an increase of fluid velocity in the axial direction or in other words results in fluid acceleration due to spatial variation in the fluid velocity.

1.6 Mass Conservation

The general equation of continuity can be obtained for an infinitesimal control volume and can be written in the differential form as

$$\frac{\partial \rho}{\partial t} + \frac{\partial(\rho u)}{\partial x} + \frac{\partial(\rho v)}{\partial y} + \frac{\partial(\rho w)}{\partial z} = 0 \tag{1.17}$$

where ρ denotes the density. Equation 1.17 is applicable under all conditions, for example, compressible or incompressible flow, viscous or inviscid, steady or unsteady flow, with or without heat transfer, etc. Density does not vary either in space or with time in an incompressible flow and therefore density gets eliminated from Eq. 1.17. Thus the simplified form of the continuity equation for an incompressible flow can be written as

$$\frac{\partial u}{\partial x} + \frac{\partial v}{\partial y} + \frac{\partial w}{\partial z} = 0 \tag{1.18}$$

However, if the flow is compressible and steady then the density gets eliminated from the unsteady term only and the simplified form of the continuity equation for a steady, compressible flow can be written as

$$\frac{\partial(\rho u)}{\partial x} + \frac{\partial(\rho v)}{\partial y} + \frac{\partial(\rho w)}{\partial z} = 0 \tag{1.19}$$

Equation 1.19 is more complex compared to Eq. 1.18 because we need to provide the spatial variation of density as an input in order to solve Eq. 1.19.

1.7 Conservation of Linear Momentum

According to the conservation of linear momentum, the vector sum of all forces, including body forces and surface forces (which include pressure forces and viscous forces) is proportional to the acceleration. Momentum is a vector quantity and accordingly differential equations for the conservation of the momentum can be written in the three directions in the Cartesian co-ordinate system as

$$\rho\left(\frac{\partial u}{\partial t} + u\frac{\partial u}{\partial x} + v\frac{\partial u}{\partial y} + w\frac{\partial u}{\partial z}\right) = \rho g_x - \frac{\partial p}{\partial x} + \frac{\partial \tau_{xx}}{\partial x} + \frac{\partial \tau_{yx}}{\partial y} + \frac{\partial \tau_{zx}}{\partial z} \tag{1.20}$$

$$\rho\left(\frac{\partial v}{\partial t} + u\frac{\partial v}{\partial x} + v\frac{\partial v}{\partial y} + w\frac{\partial v}{\partial z}\right) = \rho g_y - \frac{\partial p}{\partial y} + \frac{\partial \tau_{yx}}{\partial x} + \frac{\partial \tau_{yy}}{\partial y} + \frac{\partial \tau_{yz}}{\partial z} \tag{1.21}$$

$$\rho\left(\frac{\partial w}{\partial t} + u\frac{\partial w}{\partial x} + v\frac{\partial w}{\partial y} + w\frac{\partial w}{\partial z}\right) = \rho g_y - \frac{\partial p}{\partial y} + \frac{\partial \tau_{zx}}{\partial x} + \frac{\partial \tau_{zy}}{\partial y} + \frac{\partial \tau_{zz}}{\partial z} \tag{1.22}$$

The LHS of Eqs. 1.20–1.22 denotes the acceleration, the first term on the RHS of Eqs. 1.20–1.22 represents the body force, the second term the forces due to spatial variation in pressure and the last three terms on the RHS represent the viscous forces. It is clear from Eqs. 1.20–1.22 that the velocity field is not dependent on the magnitude of the pressure and stresses but on their gradients only.

1.8 Navier–Stokes Equations

Navier–Stokes equations can be obtained by substituting the stress terms (Eqs. 1.12, 1.13) in Eqs. 1.20–1.22 for the conservation of the linear momentum and invoking the continuity equation. The resulting Navier–Stokes equations can be written for three directions in the Cartesian coordinate system as

$$\rho\frac{\mathrm{d}u}{\mathrm{d}t} = -\frac{\partial p}{\partial x} + \mu\left(\frac{\partial^2 u}{\partial x^2} + \frac{\partial^2 u}{\partial y^2} + \frac{\partial^2 u}{\partial z^2}\right) + \rho g_x \tag{1.23}$$

$$\rho\frac{\mathrm{d}v}{\mathrm{d}t} = -\frac{\partial p}{\partial y} + \mu\left(\frac{\partial^2 v}{\partial x^2} + \frac{\partial^2 v}{\partial y^2} + \frac{\partial^2 v}{\partial z^2}\right) + \rho g_y \tag{1.24}$$

$$\rho\frac{\mathrm{d}w}{\mathrm{d}t} = -\frac{\partial p}{\partial z} + \mu\left(\frac{\partial^2 w}{\partial x^2} + \frac{\partial^2 w}{\partial y^2} + \frac{\partial^2 w}{\partial z^2}\right) + \rho g_z \tag{1.25}$$

The terms on LHS of Eqs. 1.23–1.25 denote components of acceleration (also called substantial or material derivative) and g_x, g_y and g_z denote the component of body forces in three co-ordinate directions, respectively. Equations 1.23–1.25 are elliptic in nature and therefore their numerical solution for a steady flow requires specifications of all variables (u, v, w and p if the flow is incompressible) at all boundaries of a particular flow domain. The boundary conditions for velocity are straightforward, whereas the pressure boundary conditions are somewhat tricky. Equations 1.23–1.25 must be combined with the continuity Eq. 1.18 to close the system of equations, i.e., to ensure that the number of equations is equal to number of unknowns.

The Navier–Stokes equations are second order non-linear partial differential equations and therefore it is difficult to obtain their general closed form solution for most flow situations. However, closed form solutions for some specific simple cases can be obtained.

For fluids with zero viscosity ($\mu = 0$) the governing Navier–Stokes equations reduce to the corresponding Euler's equations for inviscid flows. Bernoulli's equation is obtained by the integration of Euler's equation along a streamline and is applicable only in a steady, incompressible and frictionless flow. This equation relates pressure, velocity and elevation changes along a streamline. Bernoulli's equation essentially deals with the conservation of energy and is thus directly related to the first law of thermodynamics.

1.9 Conservation of Energy

The equation governing the energy balance (i) can be written in the differential form as

$$\rho C_p \frac{\mathrm{d}T}{\mathrm{d}t} = \left[\frac{\partial}{\partial x}\left(k\frac{\partial T}{\partial x}\right) + \frac{\partial}{\partial y}\left(k\frac{\partial T}{\partial y}\right) + \frac{\partial}{\partial z}\left(k\frac{\partial T}{\partial z}\right)\right] + \phi + \dot{q} \tag{1.26}$$

Here dT/dt denotes material or substantial derivative of temperature, C_p the specific heat at constant pressure and k denotes the thermal conductivity, which may vary over space, and $\dot{q}$ denotes a source or sink term, ϕ denotes the viscous dissipation, which is the rate of conversion of mechanical energy to thermal energy due to viscous effects. Energy is a scalar quantity and therefore only one equation is employed. This equation is valid for laminar flow only. In a turbulent flow, there is additional diffusion due to turbulent eddies. In the subsequent chapters we will deal with the conservation of energy for turbulent flows.

From the general form of the energy equation (1.26) a simplified form of the governing differential equation for thermal boundary layer may be written as

$$u\frac{\partial T}{\partial x} + v\frac{\partial T}{\partial y} = \alpha\frac{\partial^2 T}{\partial y^2} + \phi \tag{1.27}$$

Here T denotes the static temperature. The assumptions employed for thermal boundary layer equation (1.27) are similar to those for hydrodynamic boundary layer: a rather thin region close to the wall compared to the remaining region ($\delta_t \ll x$) where thermal gradients are significant and gradients in the streamwise direction (x) are negligible compared to those in the normal direction (y).

The last term in Eq. 1.27 arises due to the viscous dissipation. In most situations this term may be negligible compared to other terms in Eq. 1.27. The viscous dissipation term is significant for high speed motion of highly viscous fluids (for example, lubricating oils). Note that the thermal boundary layer equation needs to be solved in conjunction with the corresponding equation for the hydrodynamic boundary layer or in other words thermal field is dependent on the velocity field.

1.10 Boundary Conditions

In this section, we will consider the general boundary conditions that need to be specified to obtain flow and thermal fields within a computational domain using the governing differential equations presented in the previous sections.

1.10.1 No Slip and No Temperature Jump Conditions

The no slip condition means that the velocity of fluid in contact with a solid surface is equal to the surface velocity. This condition arises due to the viscosity of the fluid and is the result of the molecular interactions. In most cases, the surface is stationary and therefore the fluid velocity is assumed to be zero. Molecular viscosity is a fluid property.

If fluid is assumed to be inviscid, i.e., if we consider the Euler equations, then the fluid velocity at the solid surface is obtained as a part of the solution and is not specified as the boundary condition. The computed fluid velocity in an inviscid flow may have a finite value at a stationary wall and therefore the flow may experience a slip, i.e., relative velocity, at the wall. However, in an inviscid flow the fluid is not allowed to permeate through the surface or in other words, the velocity of fluid normal to the surface is always assumed to be zero. The no temperature jump condition is equivalent to the no slip condition and states that due to the finite thermal diffusivity of a fluid, its temperature is equal to the surface temperature it is in contact with.

1.10.2 Inlet and Outlet

At any inlet or outlet section usually the distribution of velocity field, pressure and temperature is specified, i.e., $\vec{V}, p, T$ are known. In some situations it is difficult to specify fluid behavior at the outlet and therefore another form of condition may be used in which the gradients of all flow variables in the flow direction are assumed to be zero. This condition is similar to the symmetry condition in which the gradients of all flow variables normal to the line of symmetry are assumed to be zero.

1.10.3 Interface

Specification of the boundary conditions at the interface between two different phases or a liquid free surface is somewhat more complex compared to that in the previous situations. The first condition that needs to be satisfied is the equality of vertical velocity (w) at the interface between two fluids if the interface is specified by a coordinate perpendicular to the x–y plane. This is termed as the kinematic condition. In addition one needs to satisfy the equality of one normal and two shear stresses at the interface. The effect of surface tension also needs to be accounted for in the equality for the normal stress at the interface. The heat flux normal to the interface must also be the same in both phases to maintain the thermal equilibrium.

1.11 Convective Heat Transfer

Heat transfer is energy in transit due to a temperature difference. Heat can be transferred by three modes: conduction, convection and radiation. All surfaces at finite temperature emit energy using electromagnetic waves. This mode of heat transfer termed as the radiation takes place without any medium. The other two modes of heat transfer namely, conduction and convection, can take place only in the presence of a medium. Heat transfer by random molecular motion (molecular diffusion) is always present in a fluid and is termed as the conduction heat transfer. However, there is an additional energy transfer due to bulk or macroscopic fluid motion called advection and the cumulative effect of the two (conduction and advection) is termed as the convective heat transfer (Fig. 1.4). Because of the no slip condition (and consequently no convection) at the wall the heat transfer is due to the random fluid motion only or due to conduction only. δ and δ_t in Fig. 1.4 denote hydrodynamic and thermal boundary layer thicknesses, respectively. In a laminar flow, relative positions of these thicknesses depend on the value of Prandtl number (Pr). For $Pr > 1$, at a particular location the hydrodynamic boundary-layer thickness is greater than the thermal boundary layer thickness and vice versa for $Pr < 1$. In case the solid surface is maintained at a constant heat flux then its temperature will continuously increase in the streamwise direction compared to other possibility in which it is maintained at a uniform temperature T_s all along its length.

Convection can be of two types depending upon the nature of the fluid motion: (a) forced convection and (b) free (or natural) convection. The flow is caused by external means, such as fan or blower, in a forced convection and in the later it takes place due to buoyancy forces, which could be due to thermal or concentration gradients within the flow. Rate of heat transfer is order of magnitude larger in a forced convection than that in a free convection. Therefore value of the convective heat transfer coefficient h is quite large in forced convection compared to that in free convection. Thus forced convection is always preferred over free convection. However, in certain situations, it may not be possible to have heat transfer by forced convection and therefore free convection may be the only mode present.

Since the convection heat transfer involves fluid motion, governing equations for the conservation of mass, momentum and energy form the basis of convective heat transfer analysis. The energy transferred by this mode of heat transfer is sensible or internal thermal energy of fluid.

The rate of heat transfer in convection heat transfer is governed by the Newton's law of cooling and is given as

$$q'' = h(T_s - T_\infty) \tag{1.28}$$

where q'' denotes the heat flux, h the local convective heat transfer coefficient, T_s and T_∞ denote the surface and ambient temperatures, respectively.

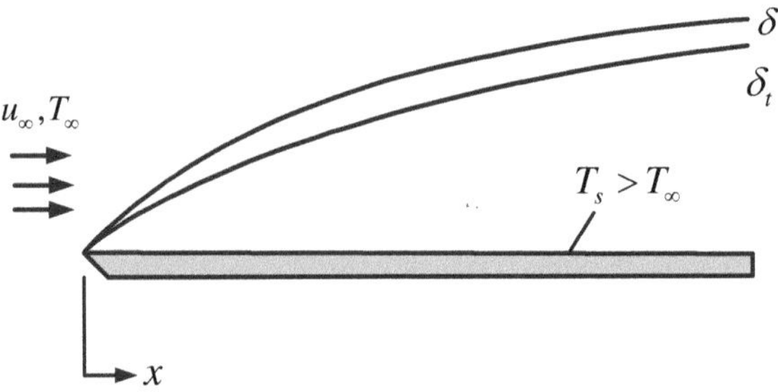

Fig. 1.4 A schematic of convective heat transfer over a heated plate

The key issue in a convective heat transfer problem is to obtain the value of h, which depends on all the factors that govern the fluid behaviour in a particular flow configuration such as the flow geometry, boundary conditions, values of physical properties, etc. The overall heat transfer coefficient can be obtained from the local value by a simple integration.

The heat flux at the surface can be obtained from the Fourier's law of conduction as

$$q_s'' = -k_f \frac{\partial T}{\partial y}\Big|_{y=0} \tag{1.29}$$

where k_f denotes the fluid thermal conductivity $\frac{\partial T}{\partial y}\big|_{y=0}$ the temperature gradient at the solid surface. By equating Eqs. 1.28 and 1.29 the local convective heat transfer coefficient can be written as

$$h = \frac{-k_f \frac{\partial T}{\partial y}\big|_{y=0}}{T_s - T_\infty} \tag{1.30}$$

Nusselt number (Nu) is an important non-dimensional number and is a measure of the convective heat transfer occurring at a surface and is equal to the non-dimensional temperature gradient at the surface

$$Nu = \frac{hL}{k_f} = \frac{\partial T^*}{\partial y^*}\Big|_{y^*=0} \tag{1.31}$$

Here L denotes a characteristic length scale, T^* and y^* non-dimensional temperature and distance from wall, respectively. Thus it is clear that the thermal behaviour in the vicinity of a solid wall governs the local heat transfer coefficient, which in turn depends on the hydrodynamic behaviour in the vicinity of the solid wall.

1.12 Examples

In this section, we will consider two examples which will illustrate the concepts presented in the preceding sections.

1.12.1 Flow Normal to an Infinite Circular Cylinder

We will consider interesting features of flow normal to an infinite circular cylinder (Fig. 1.5). The Reynolds number for this flow configuration is defined based on the

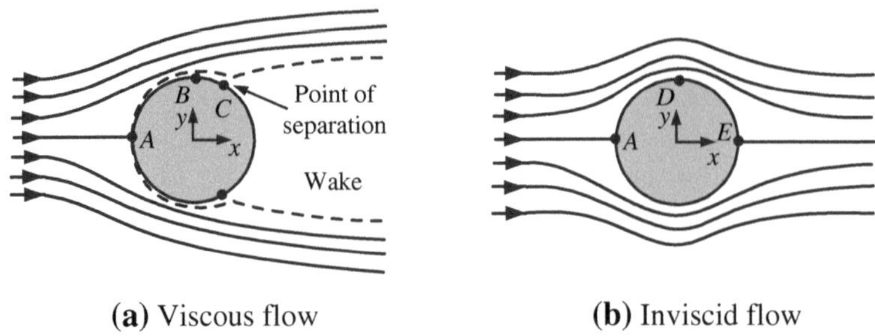

Fig. 1.5 Incompressible flow normal to a circular cylinder **a** viscous flow and **b** inviscid flow

cylinder diameter and free-stream velocity ($Re_D = U_\infty v/D$). We will consider two situations: (a) real (viscous) flow (Fig. 1.5a) and (b) ideal (inviscid) flow (Fig. 1.5b).

In a real (viscous) fluid flow past a circular cylinder a thin boundary layer begins to develop from the front stagnation point A, a point with zero velocity. The boundary-layer thickness depends on the Reynolds number of the incoming flow. A high value of Re, which means inertia dominates the viscous effects, results in a thinner boundary layer compared to that for low Reynolds number. The flow behaviour is symmetric about the x-axis, but is not symmetric about the y-axis. This is because the boundary-layer flow separates once it experiences adverse pressure gradient in the diverging passage due to the shape of the cylinder. This results in a large wake behind the cylinder (Fig. 1.5a). Flow separation causes a strong interaction between the flows in inner (close to the cylinder) and outer regions. The flow also experiences viscous normal and shear stresses resulting in net forces in the x-direction, which is termed as the drag. The amount of drag depends on the size of the wake. A larger wake results in a large pressure differential between the front and rear portions of the cylinder and therefore larger pressure drag. The flow field can be obtained from the solution of the two-dimensional Navier–Stokes equations. There is no force along with y-axis due to the symmetry of the flow about the x-axis.

In the second case the flow is symmetric about both x- and y-axes. The velocity increases along the converging portion from A to D and subsequently it decreases in the diverging passage. The variation of pressure is opposite to that of velocity or in other words it decreases in the regions where velocity increases and vice versa. The flow along the central streamline hits the cylinder at point A, which is termed as the front stagnation point (Fig. 1.5b). Likewise a rear stagnation point E exists in the flow. Since the flow is assumed to be inviscid, it has no viscous stresses and therefore we need to consider only the pressure variation to obtain the net force on the cylinder (Fox et al. 2009). Because of the symmetry of flow about both x- and y-axes the net force is zero in this case of inviscid flow. The flow field can be obtained from the solution of two-dimensional Euler's equations with suitable boundary conditions.

An interesting feature of viscous flow is that the incoming flow is always steady, whereas the flow downstream of the cylinder may be unsteady depending on the value of Re. The boundary-layer separating from the surface of the cylinder forms a free shear layer which is highly unstable. This shear layer rolls into a discrete vortex and detaches from the surface and this behaviour is called the vortex shedding. Another type of flow instability emerges as the shear layer vortices shed from both the top and bottom surfaces interact with one another. These are shed alternatively from the cylinder and generate a regular vortex pattern (the Karman vortex street) in the wake of the cylinder. The vortex shedding occurs at a discrete frequency and is a function of the Reynolds number. The dimensionless frequency of the vortex shedding, the shedding Strouhal number is defined as $St = fD/U_\infty$, where f denotes the frequency of vortex shedding.

1.12.2 Thermal Boundary-Layer Inside a Heated Circular Tube

Consider a cold fluid at temperature T_∞ axially entering a circular tube maintained at uniform temperature (T_s). Convection heat transfer takes place between the fluid and tube wall and thermal boundary-layer develops axially all around the periphery inside the tube. The thermal boundary-layers merge at a certain axial distance from the entrance. The region where this axial growth of thermal boundary-layer takes place is termed as the thermal entrance region and subsequent region is termed as the fully developed region (Fig. 1.6).

It is important to note that in the present case, the mean fluid temperature (T_m) goes on increasing downstream even in the fully developed region ($dT_m/dx > 0$) due to continuous heat transfer from heated tube surface to the cold liquid. The

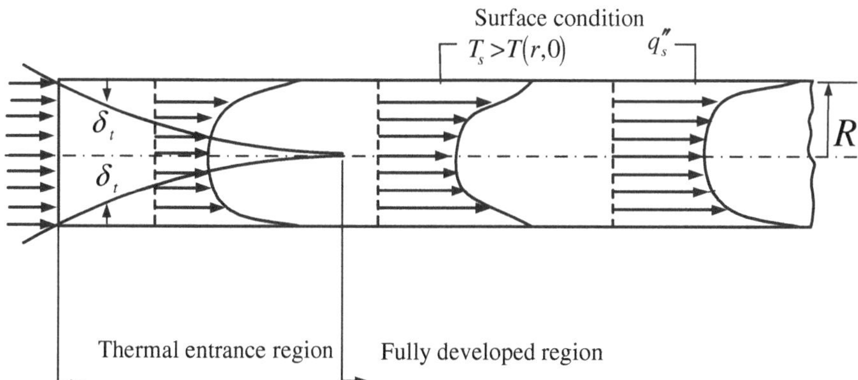

Fig. 1.6 A schematic of thermal boundary-layer inside a heated circular tube

mean temperature can be computed from the temperature profile by a simple integration as

$$T_{\mathrm{m}}(x) = \frac{2}{u_{\mathrm{m}} r_{\mathrm{o}}^2} \int_0^{r_{\mathrm{o}}} uTr\,\mathrm{d}r \qquad (1.32)$$

Here u_{m} denotes the average flow velocity (equal to the volume flow rate Q divided by the cross-sectional area) and r_{o} the inner tube radius. T_{m} is equivalent of the free-stream temperature T_{∞} in flows with the free-stream. The Newton's law of cooling in this case may be expressed as

$$q_s'' = h(T_{\mathrm{s}} - T_{\mathrm{m}}) \qquad (1.33)$$

Since the temperature profile changes continuously in the streamwise direction, we need to define the condition under which the thermal behaviour can be termed as the fully developed. It is considered to be fully developed if a non-dimensional temperature θ defined as

$$\theta = \left[\frac{T_s(x) - T(r,x)}{T_s(x) - T_m(x)} \right] \qquad (1.34)$$

does not change in the flow direction, i.e., $\partial\theta/\partial x = 0$.

It can be shown that the average Nusselt number for laminar fully developed flow through a circular tube for uniform surface heat flux condition is 4.36 and for the constant surface temperature this value is 3.66 (Incropera and DeWitt 2005).

1.13 Concluding Remarks

In this chapter, we have presented important aspects of fluid mechanics and convective heat transfer. A reader may go through the references for a detailed discussion on different aspects. In the next chapter we will introduce the reader to fluid turbulence. We will also present significance of turbulence to engineers.

References

Fox RW, McDonald AT, Pritchard PJ (2009) Introduction to fluid mechanics, 7th edn. Wiley, New York

Incropera FP, DeWitt DP (2005) Fundamentals of heat and mass transfer, 5th edn. Wiley, New York

White FM (2003) Fluid mechanics, 5th edn. McGraw-Hill, New York

Chapter 2
Fluid Turbulence

Abstract In this chapter we present characteristics of turbulent flows in general and see how a turbulent flow is so much different from laminar flow. We briefly refer to some engineering applications where turbulent fluid flow and heat transfer are encountered. We study how an already complex turbulent flow becomes much more complex due to some features such as the presence of a wall and buoyancy. We also characterize turbulent flows. We present the need to numerically study turbulent fluid flows and heat transfer and consequently what is the job of a turbulence model.

2.1 Physical Description

We encounter turbulent flows almost everywhere, for example, atmospheric and ocean currents, discharge of pollutants, oil transport in pipelines, flow through pumps and turbines, and the flow in boat wakes and around aircraft wing tips. Turbulent flows are characterized mainly by unsteadiness, vorticity, three-dimensionality, dissipation, wide spectrum of scales, and large mixing rates. In contrast, laminar flows are realized mostly in laboratories or in special situations. Fluid turbulence remains one of the biggest challenges to scientists and engineers and therefore it continues to be an active research topic.

Let us first look at the significance of some of the above-mentioned characteristics. In a turbulent flow, a flow property at a point continuously undergoes changes in magnitude. Figure 2.1 shows time traces of a typical flow property (a velocity component in the present case) that is usually obtained at a point within a turbulent flow. Turbulent flows are irregular in nature (Tennekes and Lumley 1997). This makes a deterministic approach to turbulence problems impossible and therefore statistical methods are used for treating turbulent flows. The diffusivity of turbulence causes good mixing and consequently increased rates of momentum, heat and mass transfers. Turbulent-mixing, due to the movement of eddies, is

A. Dewan, *Tackling Turbulent Flows in Engineering*,
DOI: 10.1007/978-3-642-14767-8_2, © Springer-Verlag Berlin Heidelberg 2011

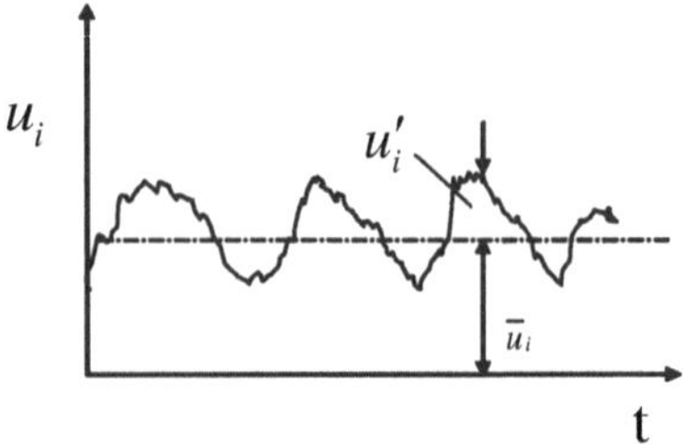

Fig. 2.1 Mean and fluctuating components of a flow property in a turbulent flow

much stronger than that of laminar flows which is due molecular action only. We must, however, not forget that mixing due to molecular effects is always present in a turbulent flow. Nevertheless, molecular mixing is negligible compared to turbulent mixing, except in certain regions (for example in the vicinity of a solid wall). Right at the solid wall, however, turbulence is zero because of the no-slip condition and therefore in the close vicinity of a solid wall the molecular effects become important. A turbulent flow is always three-dimensional and rotational.

Turbulence is characterized by high levels of fluctuating vorticity. A turbulent flow contains a wide range of eddies interacting with each other. Vortex stretching is an important mechanism in a turbulent flow. A continuous transport of energy from mean flow to large eddies, from these large eddies to a series of increasingly smaller eddies takes place and this is termed as the cascade process. The smaller eddies are influenced by the strain rate imposed by the large eddies and are continually stretched. The smallest eddies dissipate the kinetic energy into thermal energy due to viscous effects. The smallest eddy size, termed as the Kolmogorov scale, decreases with increasing Reynolds number. The largest eddy size depends on the flow configuration and is usually approximately as large as the size of the domain.

Turbulence is a continuum phenomenon, governed by the equations of fluid mechanics already described in the previous chapter. Even the smallest scales occurring in a turbulent flow are normally larger than any molecular length scale. Turbulence is not a feature of fluids but of flows. Most of the dynamics of turbulence may be similar in all fluids, whether gases or liquids or in other words independent of fluid properties.

A quantity of interest in a turbulent flow is the energy distribution over different wave numbers (or eddy sizes). If κ denotes the wave number (inversely proportional to the size of the eddy) and $E(\kappa)d\kappa$ the turbulence kinetic energy between a change of wave number $d\kappa$, then total turbulence kinetic energy may be given as

$$k = \frac{1}{2}\left(\overline{u'^2} + \overline{v'^2} + \overline{w'^2}\right) = \int_0^\infty E(\kappa)d\kappa \tag{2.1}$$

where $\overline{u'^2}, \overline{v'^2}$, and $\overline{w'^2}$ denote mean squared velocity fluctuations in three directions, respectively. $E(\kappa)d\kappa$ depends on molecular viscosity, rate of dissipation of turbulence kinetic energy (ε), integral length scale, wave number and mean strain rate. Kolmogorov (1942) proposed that for fully turbulent flows an intermediate size of eddies can be found for which the cascade process is independent of the details of the energy containing eddies and molecular viscosity. This range of eddies is termed as the inertial subrange (Fig. 2.2). Kolmogorov (1942) showed that in the inertial subrange

$$E(k) = C_k \varepsilon^{2/3} \kappa^{-5/3} \tag{2.2}$$

Here C_k denotes the Kolmogorov constant.

No general solution to the governing Navier–Stokes equations for turbulent flows is known and no general solutions to problems in turbulent flow can be found. Every turbulent flow is different, even though different turbulent flows may have many common characteristics.

Intermittency (γ) is an important parameter for turbulent flows and is defined as the fraction of time at a point for which the flow is turbulent. Obviously in a laminar flow intermittency is zero everywhere. In a turbulent flow, intermittency is also zero right at the wall because of the no slip condition. However, it rises extremely sharply close to the wall and attains the maximum value of 1.0 almost next to the wall (Fig. 2.3). The intermittency approaches zero as one moves normal to the wall and towards the free-stream.

Entrainment is another important parameter and is defined as the rate of increase of mass in the downstream direction. Entrainment and consequently the growth rate is large in a turbulent flow compared to that in a laminar flow because eddies in a turbulent flow cause outer irrotational flow to move with the main flow direction. As a result the mass flow rate of a turbulent flow increases. In almost all cases turbulence is a stable state and once the flow reaches this state, it maintains itself be supplying energy from the mean flow to replace that lost at the smallest Kolmogorov scale due to viscous effects.

We must never forget that a raw turbulent flow is always unsteady and three-dimensional. Engineers are mostly interested in time averaged flow and such flow may have many simplifications such as steady or two-dimensional and so on. Before we move on to next section, we must pause for a few minutes and try to

Fig. 2.2 Energy spectrum for a fully turbulent flow

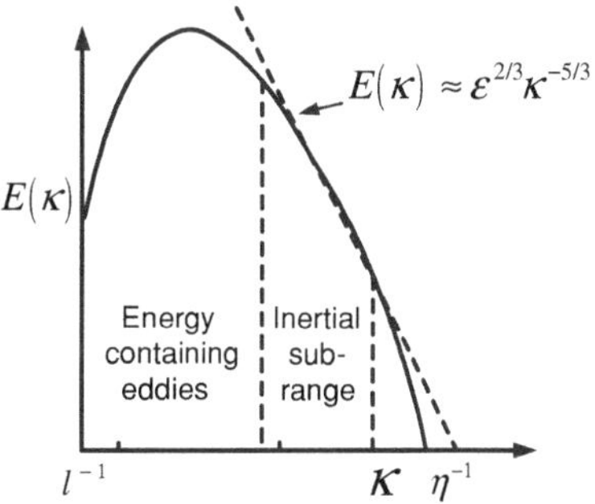

Fig. 2.3 Intermittency profile for turbulent flow past a flat plate. γ is zero right at the wall ($y = 0$), but it rises sharply to 1.0 and therefore this behaviour cannot be shown in a linear plot

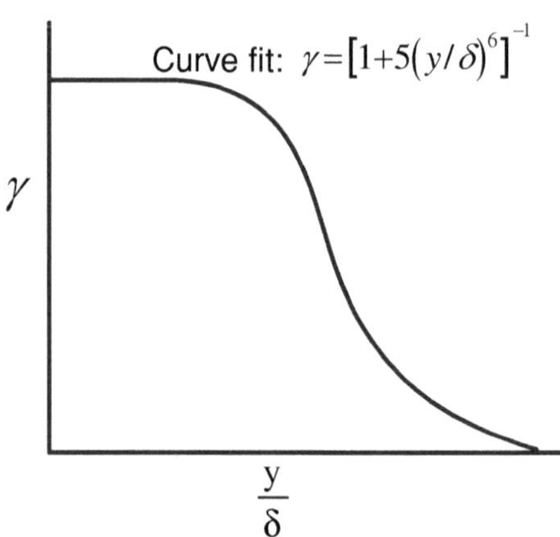

understand main differences between laminar flows and turbulent flows and appreciate the complexities and salient features of turbulent flows.

2.2 Stability of Laminar Flows

A flow becomes turbulent when a laminar flow becomes unstable at a high value of the governing parameter, for example, Reynolds number in viscous flows and Grashof number in buoyancy driven flows. The instabilities are related to a complex interaction of viscous and nonlinear terms in the governing momentum equations. Linear hydrodynamic stability theory deals with conditions that lead to the amplification of disturbances in a flow. A basic treatment of stability of a flow involves selection of a basic flow, adding a disturbance to the flow to arrive at governing equations for disturbance, linearizing the governing equations and solving for the eigenvalues to arrive at the conditions for the instability of flow. The resulting governing equations for the disturbance are termed as the Orr–Sommerfeld equations. A study of stability of laminar flows is a complex subject and separate books have been dedicated to this subject (Chandrasekhar 1961; Drazin and Reid 1996; Drazin 2002). Typically free-shear flows are quite unstable even at low values of Reynolds number (typically about 10) and readily undergo transition. Wall bounded flows are known to be more stable.

2.3 Transition and Onset of Turbulence

Engineers are primarily interested in the prediction of the values of the governing parameter, for example critical Reynolds number Re_{crit} at which disturbances occurring in the flow become unstable and transitional Reynolds number Re_{tr} at

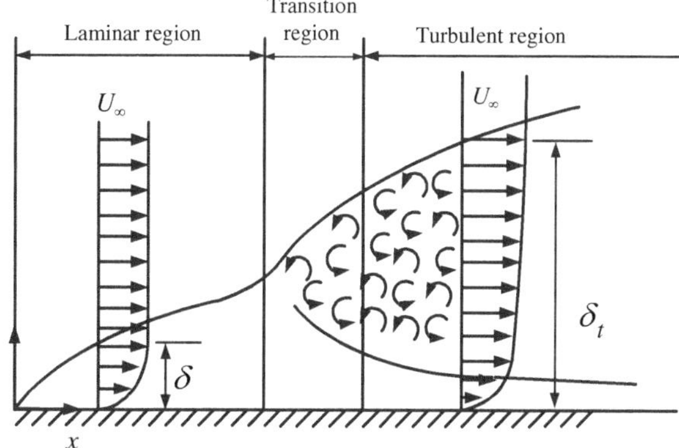

Fig. 2.4 Turbulent boundary-layer flow past a flat plate

which disturbances amplify to such an extent that they begin to influence the mean flow. The former can be predicted by the linear stability theory. However no theory exists that can predict the value of Re_{tr} for a particular flow. Some values have been proposed in the literature based on experimental data or correlations.

In a zero pressure gradient flow past a flat plate the value of $\mathrm{Re}_{x,cr}$ approximately equals 91,000 and different researchers have proposed different values of $\mathrm{Re}_{x,tr}$ with a range from 2.0×10^6 to 4.7×10^6 (Fig. 2.4). In a flow through a circular pipe, flow is always laminar for $\mathrm{Re}_D < 2,000$ and flow becomes turbulent for $\mathrm{Re}_D > 4,000$ depending on inlet conditions. Reynolds (1883) performed the famous dye experiment to study transition to turbulent flows in a circular pipe with several interesting outcomes.

The values of Reynolds number encountered in practical applications are such that flows are almost always turbulent. Most free-shear flows become unstable at very small values of Reynolds number and therefore for all practical purposes may be assumed turbulent under all conditions. The process of change in flow behaviour from laminar to turbulent flow is termed as the transition. A sequence of events in flow past a flat plate undergoing laminar to turbulent transition can be described as (White 2007): first the stable laminar flow experiences unstable two-dimensional Tollmein–Schlichting waves followed by the development of three-dimensional waves. Subsequently there is vortex breakdown locally where high shear occurs followed by three-dimensional fluctuations. This is followed by the local formation of turbulent spots and coalescence of these spots into fully turbulent flow.

2.4 Types of Turbulent Flows

- *Wall bounded and free-shear flows.* Turbulence in the presence of walls (termed as wall bounded turbulence) is quite complex compared to that far away from a wall. In Sect.3.1 in Chap.3 we will present a detailed treatment of turbulence in the presence of a solid wall.

- *Homogeneous turbulence.* If fluid turbulence has the same behaviour in all parts of the flow field, then it is termed as the homogenous turbulence.
- *Isotropic turbulence.* If the statistical features of turbulence are same in all flow directions, then it is termed as the isotropic turbulence.
- *Stationary turbulence.* If the time-average of all flow properties in a turbulent flow does not vary with time, then it is termed as the stationary turbulence.

It may be noted that in general no turbulent flow is homogenous or isotropic.

2.5 Significance of Turbulent Flows and Heat Transfer

We need to device ways to model/simulate turbulent flows that are encountered in a wide range of engineering applications. With the emergence of computers and different computational techniques, the computational fluid dynamics (CFD) has become an important tool for investigating turbulent flows. In CFD a set of appropriate governing equations are solved numerically within a computational domain using appropriate boundary conditions to obtain flow behaviour. The present book the fluctuations in the density are not considered, or in other words the treatment of the present book is applicable only for incompressible flows. Turbulence is usually specified as a percentage of mean flow properties $T_i = \sqrt{(2/3)k}/U_{ref}$ where k denotes turbulence kinetic energy and U_{ref} a reference velocity. This expression is based on the assumption of isotropic turbulence, i.e., $\overline{u'^2} = \overline{v'^2} = \overline{w'^2}$.

Let us consider the complexity involved in computing a turbulent flow. A typical domain of 0.1 m by 1.0 m with a high Reynolds number turbulent flow might contain eddies down to 10–100 μm sizes and therefore the computing meshes of 10^9 to 10^{12} points may be required to accurately describe the processes at all length scales using the numerical solution of unsteady, three-dimensional Navier–Stokes equations. The fastest events may take place with a very large frequency of the order of 10 kHz and therefore we need to march in time with extremely small steps (of the order of hundreds of micro seconds). Thus we require extremely large grid points in both three dimensions and time. This approach of numerically solving unsteady, three-dimensional Navier–Stokes equations without any approximations or assumptions and resolving all scales in space and time is termed as the direct numerical simulation (DNS). Since a direct solution of the time-dependent Navier–Stokes equations of turbulent flows is computationally expensive, it is restricted to low Reynolds numbers and simple geometries. Fortunately, in many situations only time-averaged flow properties are required by engineers and thus there may not be a need to obtain an instantaneous numerical solution.

We will devote a large part of the present book on ways to compute time averaged flow properties such as velocity and temperature profiles, heat transfer coefficient, pumping power, and drag, etc. Subsequently, we will treat issues related to DNS of turbulent flows. We will also consider large eddy simulation of turbulent flows which is somewhat similar to DNS. In LES, three-dimensional, Navier–Stokes equations are numerically solved for instantaneous flows. However, one does not need to go down to the smallest Kolmogorov scale as in a DNS. The effect of smallest scales is modeled by means of subgrid scales and such models are termed as the subgrid scale models. The computations requirements for LES are somewhat less than those for DNS. LES has good potential for the design applications.

Now we will briefly consider three practical situations and see how turbulence is important in these situations. The flow within a stirred vessel that is used in several industries is highly turbulent, three-dimensional and unsteady. CFD requires that the computational grid match the shape of the vessel and all its components, which are often geometrically complex. The grid chosen should be fine enough to capture the smallest turbulent flow scales. A very fine grid will result in large computational requirements.

The design procedure of heat exchangers is quite complicated, as it involves an analysis of heat transfer rate and pressure drop. The major challenges to the design of a heat exchanger are to make it compact, which is, to achieve a high heat transfer rate and at the same time to allow its operation with a small power loss (or pumping power). The flow is usually turbulent in a heat exchanger and modeling of turbulent fluid flow and heat transfer becomes an important issue in the design of heat exchangers.

A plume is the simplest example of buoyancy driven flows, and it has important applications in environmental studies, atmospheric science and many engineering applications such as the cooling of electronic equipment, waste disposal in the atmosphere and oceans. An understanding of the plume behaviour is important for modeling more complicated buoyancy driven flows such as buoyant jets. In addition, plume provides a good example for illustrating the influence of buoyancy on turbulence. A turbulent plume is significantly more complex compared to its non-buoyant counterpart (a jet) because of interplay between buoyancy and turbulence and as we will see in Sect.9.4 in Chap. 9 that the modeling of a turbulent plume is a challenging task.

2.6 Turbulence in the Vicinity of a Solid Wall

Turbulence in the presence of solid walls exhibits interesting flow features. The region near the wall can be divided in three distinct layers: (a) viscous sublayer, where molecular effects dominate; (b) buffer layer, where the molecular and turbulent properties of the flow are both important and (c) fully turbulent layer, where the turbulent properties of the flow play the major role and molecular effects may be ignored. What is the influence of wall in a turbulent flow? The no-slip condition at the wall means the fluid velocity at the wall is equal to the wall velocity that in most cases is zero. As a result the viscous damping and kinematic blocking near the wall reduce the velocity fluctuations to zero. However, extremely large gradients in flow properties occur near the wall and therefore it can be said that walls cannot resist turbulence.

Figure 2.5 shows the profile of average streamwise velocity in the wall co-ordinates in the vicinity of a solid wall. Here $u^+ = u/u_\tau$ and $y^+ = yu_\tau/v$ denote the non-dimensional velocity and normal distance from the wall, respectively. u_τ denotes the friction velocity and is given as

$$u_\tau = \sqrt{\frac{\tau_w}{\rho}} \tag{2.3}$$

Here τ_w denotes the wall shear stress. In the innermost viscous layer from $y^+ = 0$ to $y^+ = 5$ the non-dimensional velocity profile varies linearly

$$u^+ = y^+ \tag{2.4}$$

From y^+ equal to 30 to nearly $y/\delta = 0.1$ termed as the overlap layer or logarithmic layer the profile follows the logarithmic behaviour

$$u^+ = \frac{1}{\kappa}\ln y^+ + B \tag{2.5}$$

Here κ is the von Karman constant $= 0.41$. The profile in the outer region y/δ from 0.1 to 1.0 depends on the pressure gradient and Reynolds number. It may be noted that for y^+ from 5 to 30 a smooth transition between the linear behaviour and logarithmic behaviour is observed.

In the separation region the skin friction (and therefore the friction velocity) approaches zero and therefore u^+ approaches infinity and y^+ approaches zero. For a

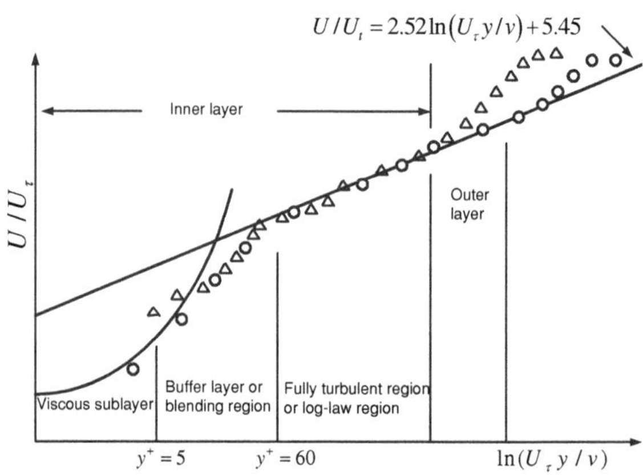

Fig. 2.5 Velocity profile in the wall coordinates in the presence of a solid wall

separating flow, the logarithmic behaviour does not hold and no overlap layer is observed.

The nature of the logarithmic overlap layer for a turbulent flow depends on the roughness of the solid wall. A small amount of roughness can disturb an extremely thin viscous sublayer and significantly increase the wall friction. The intercept B in Eq. 2.5 moves downward with an increasing roughness. The amount of shift is different at different non-dimensional values of roughness. The consequence of this shift will be discussed when we consider friction factor using the Moody's diagram in Sect. 3.5 in Chap.3.

Mean temperature in the wall co-ordinates also behaves similar to the mean velocity (Fig. 2.6). Right in the vicinity of the solid wall the mean temperature profile is linear in the thermal sublayer

$$T^+ = \Pr y^+, \tag{2.6}$$

where Pr denotes the Prandtl number and T^+ denotes the mean temperature in wall coordinates and is defined as

$$T^+ = \frac{T_w - \overline{T}}{T^*} \text{ where } T^* = \frac{q_w}{\rho C_p u_\tau} \tag{2.7}$$

Further away a logarithmic behaviour is observed.

$$T^+ = \frac{\Pr_t}{k} \ln y^+ + A(\Pr) \tag{2.8}$$

Here T^* denotes the wall conduction temperature and is equivalent of the friction velocity and $\overline{T}$ denotes the mean temperature. The last term in Eq. 2.8 is a strong function of Prandtl number (Pr).

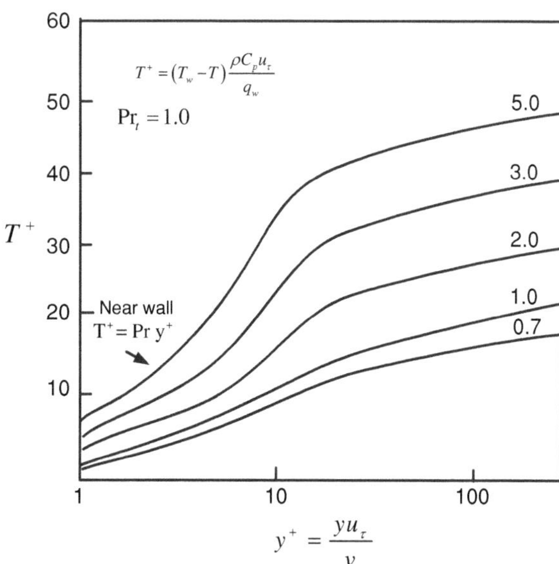

Fig. 2.6 Mean temperature profile in the wall coordinates in the presence of a solid wall

Kader (1981) provides a good curve fit to $A(\text{Pr})$ that works well for a wide range of Prandtl numbers.

$$A(\text{Pr}) = (3.85\,\text{Pr}^{1/3} - 1.3)^2 + 2.12\ln(\text{Pr}) \tag{2.9}$$

Some researchers have proposed curve fits for complete temperature profile covering sublayer, overlap layer and outer layer for a range of pressure gradients. Recently Cruz and Silva-Freire (2002) have proposed a thermal law of the wall for separating and reattaching flows.

The turbulent Prandtl number is defined as

$$\text{Pr}_t = \frac{C_p \mu_t}{k_t} \tag{2.10}$$

where C_p denotes the specific heat at constant pressure and k_t thermal diffusivity of fluid due to turbulence.

The first question that arises is, "what value of turbulent Prandtl number Pr_t should be used?". For $\text{Pr} < 0.7$ the value of Pr_t decreases from approximately 1.5 in the viscous sublayer and the adjoining region to approximately 1.0 in the logarithmic layer (Blackwell 1973). Therefore a constant value of $\text{Pr}_t \sim 1.0$ can be assumed in the computations. A large value of Pr_t in close vicinity of a solid wall is not of much significance because turbulence is anyway negligible in this region. For low values of Prandtl number (Pr) the turbulent Prandtl number seems to have a higher value (between 2 and 4). For turbulent free-shear flows a constant value of Pr_t equal to 0.7 may be used everywhere.

2.7 Task of a Turbulence Model

Having seen complexities associated with fluid turbulence we may wonder how one can model such flows. Fortunately, there are ways to predict such flows using different turbulence models. However, no turbulence model seems to be universally applicable under a wide variety of applications and therefore it is important to choose a suitable turbulence model for a particular situation. In the subsequent chapters we will see salient features of important turbulent flows. We will also see that one needs to make several assumptions to mathematically model and compute complex turbulent flows. Therefore to gain confidence in a particular turbulence model, the predictions need to be first validated for a flow configuration by comparing the predictions with experimental data or DNS data. Once the model is validated it can be applied to a new flow configuration with the hope that it will work well in a new flow configuration as well. One must realize that due to inherent several complexities in a turbulent flow, it is quite likely that a single turbulence model will not be able to handle reasonably well a wide variety of turbulent flows. We will study a wide variety of turbulence models and it will not be easy to conclude which of the models studied will be suitable for a particular flow situation.

2.8 Concluding Remarks

In this chapter we have introduced a reader to turbulence and have seen key features of turbulent flows. We have seen how a laminar flow is different from a turbulent flow and some factors that lead to further complexity in a turbulent flow. Engineers need to understand turbulent flows in order to control them. Therefore prediction is an important component of engineering calculations. In the next five chapters we will examine different ways that can be adopted to predict/simulate turbulent flows.

References

Blackwell BF (1973) The turbulent boundary layer on a porous plate: an experimental study of the heat transfer behaviour with adverse pressure gradients, Ph.D thesis. Stanford University, USA

Chandrasekhar S (1961) Hydrodynamic and hydromagnetic stability. Oxford University Press, London

Cruz DOA, Silva-Freire AP (2002) Note on a thermal law of the wall for separating and recirculating flows. Int J Heat Mass Transf 45(7):1459–1465

Drazin PG (2002) Introduction to hydrodynamic stability. Cambridge University Press, New York

Drazin PG, Reid WH (1996) Hydrodynamic stability. Cambridge University Press, London

Kader BA (1981) Temperature and concentration profiles in fully turbulent boundary layers. Int J Heat Mass Transf 24(9):1541–1544

Kolmogorov AN (1942) Equations of turbulent motion of an incompressible fluid. Izvestia Acad Sci USSR 6(1, 2):299–303

Reynolds O (1883) On the experimental investigation of the circumstances which determine whether the motion of water shell be direct or sinuous, and the law of resistance in parallel channels. Philos Trans R Soc Lond Ser A 174:935–982

Tennekes A, Lumley JL (1997) A first course in turbulence. MIT Press, USA

White FM (2007) Viscous fluid flow, 3rd edn. McGraw-Hill, New York

Chapter 3
Characteristics of Some Important Turbulent Flows

Abstract In this chapter we present characteristics of some important specific turbulent flows, which include wall bounded flows, free shear flows, separated flows, flows driven by buoyancy, turbulent convective heat transfer past a heated flat plate. We introduce the basic concepts and governing equations for studying such flows. These flows may be used as a building block for investigating more complex flows of engineering interest.

3.1 Boundary-Layer Flow Past a Flat Plate

If the velocity profile for a particular flow does not have a point of inflexion, then the viscous instability theory predicts that there is a finite region of Reynolds number approximately equal to $Re_\delta = 1,000$ (here δ denotes the boundary-layer thickness) at which the infinitesimal disturbances are amplified. Experiments substantiate these theoretical predictions that if the incoming flow is laminar then initial linear instability occurs at $Re_{x,\text{crit}}$ approximately equal to 91,000. The unstable two-dimensional disturbances are called Tollmien-Schlichting (T-S) waves and are amplified in the flow direction. Linear amplification takes place over a limited range of Reynolds numbers. If the amplitude is large enough, a secondary, non-linear, instability mechanism causes the T-S waves to become three-dimensional and eventually evolve into hairpin vortices. Transition of a natural flat-plate boundary-layer involves the formation of turbulent spots at active sites and the subsequent merging of different turbulent spots convected downstream by the flow. This takes place at Reynolds numbers $Re_{x,\text{tr}} = 10^6$.

The turbulent boundary-layer adjacent to a solid surface consists of two regions. In such flows the effects of inertia are dominant in a rather large normal region away from the wall and the behavior in this region is not affected by the molecular viscosity. A thin layer close to the wall also exists where viscous effects are important. However, in the outer region only the effects due to turbulence are

A. Dewan, *Tackling Turbulent Flows in Engineering*,
DOI: 10.1007/978-3-642-14767-8_3, © Springer-Verlag Berlin Heidelberg 2011

important. An intermediate region also exists where viscous and turbulent stresses are both important, but turbulent stresses are more than viscous stresses.

At the solid surface the fluid is stationary due to the no-slip condition and therefore turbulent fluctuations due to motions of eddies are zero at the wall. In the absence of turbulent shear stress effects the fluid closest to the wall is dominated by viscous shear only.

As we have already mentioned, turbulence in the presence of a solid wall exhibits several interesting features: (a) Flow is highly anisotropic up to y/δ approximately equal to 0.8 with root mean square turbulent fluctuations in three directions being unequal. The flow becomes isotropic at $y/\delta > 0.8$; (b) Peaks in turbulent quantities are observed extremely close to the wall though turbulence is zero right at the wall thus causing sharp gradients at the wall; (c) The streamwise velocity fluctuations are the largest, approximately equal to 10% of the mean velocity. The fluctuations normal to the wall are the smallest of the three and the ones in the spanwise direction have the intermediate values; and (d) An outer instantaneous super layer separates the inner turbulent region from the outer non-turbulent region (Klebanoff 1955). The extent of this super layer varies from approximately 0.4δ to 1.2δ and it propagates approximately at a speed of $0.9U$ in the stream-wise direction. This super layer causes the flow to be intermittent even at the edge of the boundary-layer.

Now we will briefly consider the concept of thermal boundary layer and convective heat transfer. Figure 3.1 shows flow of a cold fluid past a heated flat plate. As a result both the hydrodynamic and thermal boundary-layers begin to develop from the leading edge. The cold fluid flowing over the heated plate gains temperature by convective heat transfer from the heated plate. Due to the no slip and no temperature jump conditions at the wall, the fluid is stationary and the temperature of the fluid is equal to the local temperature of the plate.

The boundary-layer thickness δ is defined as the local distance normal to the wall where the local streamwise velocity is 99% of the freestream velocity. The other two important thicknesses relevant to the boundary layer are the displacement thickness (δ^*) and momentum thickness (θ). The former is a measure of the mass deficit and the later that of momentum deficit. The two thicknesses are given as

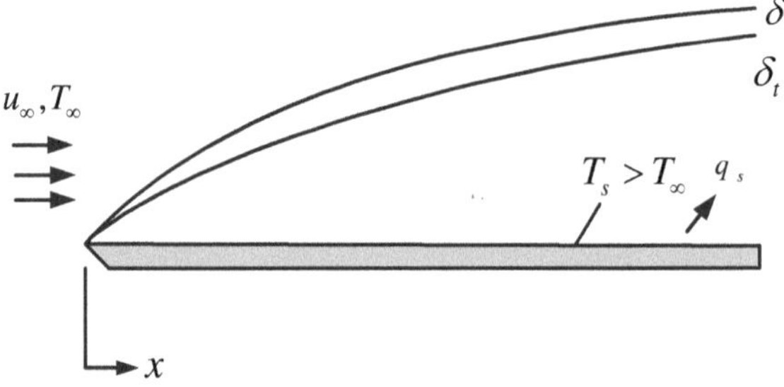

Fig. 3.1 A schematic of a thermal boundary layer (either the constant wall temperature T_s or constant heat flux condition q_s is used)

$$\theta = \int_0^\infty \frac{u}{U_\infty}\left(1 - \frac{u}{U_\infty}\right) dy \quad \text{and} \quad \delta^* = \int_0^\infty \left(1 - \frac{u}{U_\infty}\right) dy \tag{3.1}$$

where U_∞ denotes the free-stream velocity. The shape factor is defined as $H = \delta^*/\theta$. A value of H close of one implies a steeper profiles compared to those for higher values of H.

We can perform a simple integral analysis of boundary-layer flow past a flat plate for both laminar and turbulent situations. We need to prescribe an appropriate velocity profile for the two cases and obtain the variation of the drag coefficient (C_D), skin friction coefficient (C_f), boundary-layer thickness (δ), displacement thickness (δ^*) and momentum thickness (θ) along the flow direction. Here C_f and C_D are defined as

$$C_f = \frac{\tau_w}{1/2\rho U_\infty^2} \quad \text{and} \quad C_D = \frac{D}{1/2\rho U_\infty^2 A} \tag{3.2}$$

Here τ_w denotes the wall shear stress $\tau_w = \mu\frac{\partial u}{\partial y}\big|_{y=0}$, D the total drag force and A the projected area.

The use of the parabolic velocity profile for a laminar boundary-layer given by

$$\frac{u}{U_\infty} = \left(2\frac{y}{\delta} - \frac{y^2}{\delta^2}\right) \tag{3.3}$$

results in the following variations of the important properties

$$\frac{\delta}{x} = \frac{5.5}{\sqrt{Re_x}}, \quad \frac{\delta^*}{x} = \frac{1.83}{\sqrt{Re_x}} \quad \text{and} \quad C_f = \frac{0.73}{\sqrt{Re_x}} \tag{3.4}$$

Likewise we can use the following 1/7th power law for the turbulent velocity profile

$$\frac{u}{U_\infty} = \left(\frac{y}{\delta}\right)^{1/7} \tag{3.5}$$

to obtain the corresponding relations for a turbulent flow

$$\frac{\delta}{x} = \frac{0.16}{Re_x^{1/7}}, \quad \delta^* = \frac{\delta}{8} \quad \text{and} \quad C_f = \frac{0.027}{Re_x^{1/7}} \tag{3.6}$$

A comparison of Eqs. 3.4 and 3.6 shows that boundary layer grows rapidly in a turbulent flow and skin friction coefficient decays rapidly in a turbulent flow compared to that in a laminar flow.

The governing equations for the two-dimensional, mean, turbulent, boundary-layer flow such as the one shown in Fig. 3.1 may be written as

$$\text{Continuity:} \quad \frac{\partial \overline{u}}{\partial x} + \frac{\partial \overline{v}}{\partial y} = 0 \tag{3.7}$$

$$\text{Streamwise momentum:} \quad \overline{u}\frac{\partial \overline{u}}{\partial x} + \overline{v}\frac{\partial \overline{u}}{\partial y} = U_\infty \frac{dU_\infty}{dx} + \frac{1}{\rho}\frac{\partial \tau}{\partial y} \tag{3.8}$$

$$\text{Thermal energy:} \quad \rho c_p \left(\overline{u}\frac{\partial \overline{T}}{\partial x} + \overline{v}\frac{\partial \overline{T}}{\partial y} \right) = \frac{\partial q}{\partial y} + \tau \frac{\partial \overline{u}}{\partial y} \tag{3.9}$$

where τ and q denote the total shear stress and total heat flux, respectively, and are defined as

$$\tau = \mu \frac{\partial \overline{u}}{\partial y} - \rho \overline{u'v'} \tag{3.10}$$

$$q = k \frac{\partial \overline{T}}{\partial y} - \rho c_p \overline{v'T'} \tag{3.11}$$

where $\rho\overline{u'v'}$ denotes Reynolds shear stress and $\rho c_p\overline{v'T'}$ turbulent heat flux in the normal direction. These two terms are an outcome of the time-averaging of the governing momentum and energy equations described in detail in Sect. 4.2 in Chap. 4. In a laminar flow these two terms are zero.

Since we do not have an equation for the momentum in the normal direction the corresponding velocity component (v) is computed from the continuity equation (3.7). The momentum and thermal energy equations (3.8) and (3.9) are parabolic in nature and are simpler compared to the Navier–Stokes equations that are elliptic in nature. The boundary-layer equations are valid for large values of Reynolds number.

Blasius (1908) showed that for a laminar boundary-layer the velocity profile in the non-dimensional form u/U for a hydrodynamic boundary-layer in a zero pressure gradient is a function of a single dimensionless variable $\eta = y\sqrt{U/xv}$. As a result the governing boundary layer equations reduce to the ordinary differential equations (termed as the Blasius equations) which can be solved by performing the simple numerical integration. The solution provides the streamwise variations of the boundary layer thickness, skin friction coefficient (consequently drag) that are found to be in good agreement with the experimental data (Liepmann 1943).

For a turbulent boundary layer, it can be shown that though there is a small pressure variation due to turbulent velocity fluctuations, the overall pressure variation across the flow may be assumed to be negligible (White 2006).

3.2 Forced and Free Convections

Convection denotes energy transfer between a surface and fluid moving over it when both are maintained at different temperatures. Energy is transferred by molecular diffusion from the wall to the first fluid layer and then it is advected by the bulk motion of fluid (Incropera and DeWitt 2005). The development of thermal boundary-layer is similar to that of hydrodynamic boundary-layer. The ratio of hydrodynamic and thermal boundary layer thicknesses depends on the value of Prandtl number in laminar flow and turbulent Prandtl number for a turbulent flow.

The flow is caused by an external means in a forced convection, whereas it is due to buoyancy (body) forces in a free (or natural) convection. The rate of heat transfer is quite large in forced convection, but this is at the cost of energy consumption to drive the external device. There can be situations in which the only possible way to transfer heat may be free convection only and therefore this mode though involving small rate of heat transfer may be important.

Complexity involved in the computation of turbulent forced convection is slightly more than that for turbulent hydrodynamic boundary-layer. We can first compute the mean flow field using a suitable turbulence model and then temperature field may be obtained as a passive scalar. The turbulent heat fluxes may be modeled by using the concept of eddy viscosity and by using a constant value of turbulent Prandtl number (explained in Sect. 2.6 in Chap.2). However, the situation is far more complex in a turbulent free-convection with the appearance of buoyancy term in the momentum equation (Fig. 3.2). The buoyancy term in turn may depend on either temperature field or concentration field and therefore momentum equations and thermal energy equation get coupled or in other words flow field cannot be obtained independent of the temperature field. The complexity is compounded due to the production of turbulence due to buoyancy in addition to the production due to shear. We will see these details when we deal with modeling of turbulent round plume in Sect. 9.4 in Chap.9.

3.3 Simple Free Shear Flows

These flows are not affected by the presence of walls and therefore are less complex compared to wall bounded flows. We will briefly consider four types of free shear flows, namely, jet, wake, mixing layer and plume (Fig. 3.3). Far

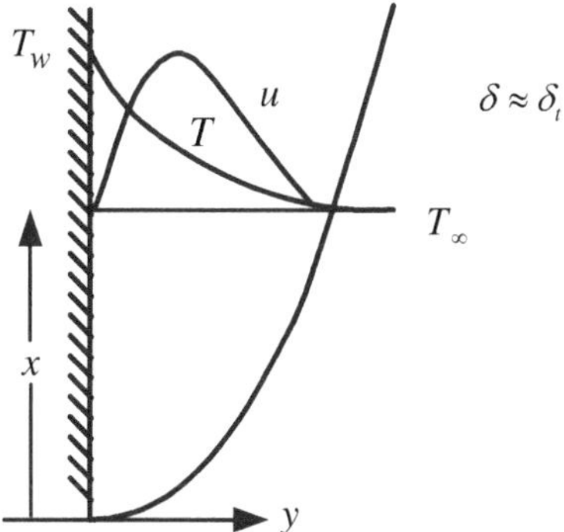

Fig. 3.2 An example of free convection due to a heated vertical plate

Fig. 3.3 A schematic
of different types of free
shear flows. **a** Jets, **b** mixing
layer, **c** wake and **d** plume

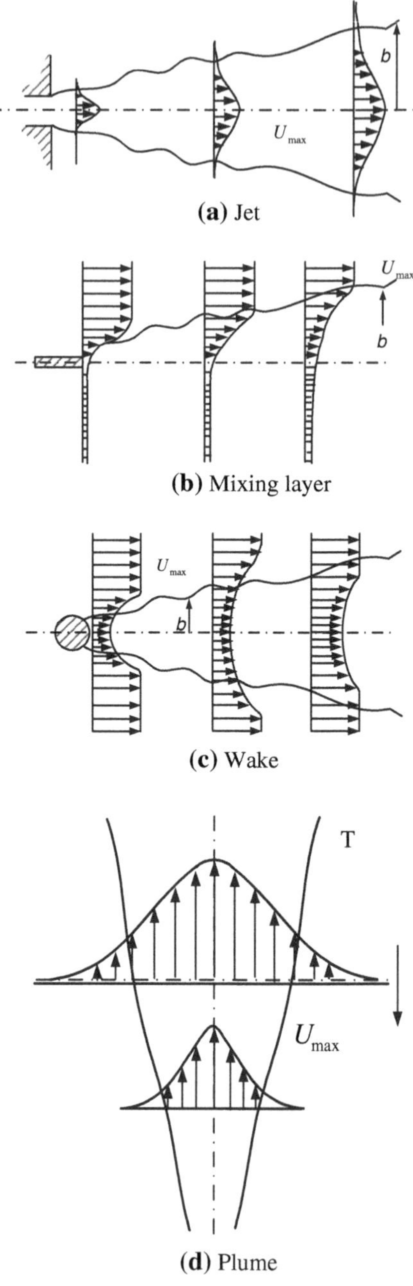

downstream these flows become self similar which means the flow profiles have
the same form if these are non-dimensionalized by using suitable variables, i.e., far
downstream their behaviour depends on local flow properties and not on flow

history. These flows grow in lateral extent due to the entrainment of the surrounding fluid. Under normal conditions and for high values of Reynolds number, the boundary-layer equations can be applied to such flows thus simplifying the analysis. It may be noted that profiles presented in Fig. 3.3 correspond to the time-averaged flow. The instantaneous profiles of these flows will have no resemblance to the time averaged flow.

A jet is formed when fluid issues out of an opening (Fig. 3.3a). Depending on type of opening the jet may have different cross sections, circular, plane, or rectangular, etc. Jets are encountered in many practical applications. Pressure remains constant in a jet and as a result the momentum of a jet remains constant and equal to the discharge momentum. The mass flow rate increases downstream due to the flow entrainment. Because of turbulent eddies, entrainment is high in a turbulent jet compared to that in a laminar jet. The local half width of a jet is defined as the distance from the axis or plane of symmetry to the normal or radial location where the local velocity is half of the local centerline velocity.

A mixing layer is formed at the interface of two streams moving at different velocities (Fig. 3.3b). At the first meeting point between the two streams there is a discontinuity in the velocity and further downstream the discontinuity vanishes due to the turbulent diffusion resulting in a smooth velocity distribution. A characteristic width of a mixing layer is defined as the difference between the values of y/x where $(U - U_2)^2/(U_1 - U_2)^2$ is between 9/10 and 1/10. Here U_1 and U_2 denote the velocities of two layers. Velocity changes across an initially thin layer are important. Transition to turbulence occurs after a very short distance in the flow direction from the point where the different streams initially meet, the turbulence causes good mixing of adjacent fluid layers and quick widening of the region across which the velocity changes take place. Initially fast moving jet fluid will lose momentum to speed up the stationary (or slowly moving) surrounding fluid. Due to the entrainment of the surrounding fluid the velocity gradient decreases in magnitude in the flow direction. These gradients disappear completely at infinity in the flow direction.

A wake is formed behind an object when it comes in the path of fluid stream. It is basically a defect in the velocity (Fig. 3.3c). A characteristic width of a wake is defined as the radial location from the centreline where the velocity defect is half of the maximum velocity defect, which typically occurs at the centreline.

A plume is a kind of jet flow that is driven by buoyancy only (Fig. 3.3d). Turbulent plumes are quite complex compared to turbulent jets because in the former the buoyancy affects both mean and turbulent quantities. Due to the additional production of turbulence due to buoyancy in a plume the growth rate (and the rate of entrainment of the surrounding fluid) is larger than that in a jet. The half width of a plume is defined in the same way as for the round jet.

3.4 Circular Pipe and Parallel Plates

A fully developed flow through a circular pipe occurs when the lateral dimensions of the viscous regions formed due to the no slip condition at the periphery of pipe all around the axis increase along the streamwise direction and eventually these merge at the axis from all sides. No variation in the flow properties along the streamwise direction takes place further downstream (Fig. 3.4).

In practice, transition to turbulence takes place in a circular pipe between Re ($=U_{av}D/v$) 2,000 and 10^5 depending on the inflow conditions. As in flat plate boundary layer, turbulent spots appear in the near wall region of a pipe flow. These grow, merge and finally fill the pipe cross-section to form turbulent spot. The transition to turbulence in a circular pipe is strongly affected by factors such as pressure gradient, disturbance levels, wall roughness and heat transfer.

Figure 3.5 shows a comparison of the velocity profiles for the same mass flow rates (and consequently for the same volume flow rate in an incompressible flow) for the fully developed laminar and time-averaged fully developed turbulent flow through a circular pipe. Laminar velocity profile is parabolic and can be derived from the governing momentum equation using the condition of flow symmetry as

$$u = u_{max}(1 - r^2/R^2) \tag{3.12}$$

where R denotes the pipe radius and r the radial distance from the axis. u_{max} denotes the maximum velocity at the centerline and is given as

$$u_{max} = (-dp/dx)(R^2/4\mu) \tag{3.13}$$

Flow in a circular pipe is driven by the favourable pressure gradient $dp/dx < 0$ and hence the negative sign in Eq. 3.13. The average velocity for a laminar flow through a circular pipe is one half of the maximum (centerline) velocity. The wall shear stress is equal to $(R/2)(\Delta p/L)$ where Δp denotes the pressure drop over length L of the pipe. An exact relation between the friction factor (and consequently pumping power) and Reynolds number for the fully developed laminar flow

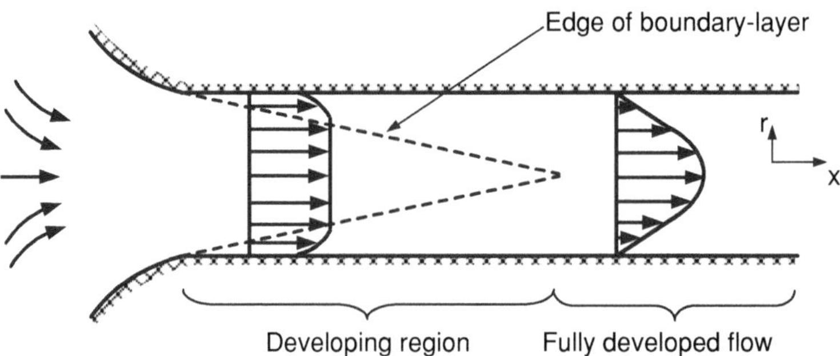

Fig. 3.4 Flow development in a circular pipe

through the circular pipe can be derived from the momentum equation and is given as

$$f = 64/\mathrm{Re_d} \tag{3.14}$$

where f denotes the Darcy friction factor given as

$$f = 8\tau_w/\rho U_{\mathrm{avg}}^2 \tag{3.15}$$

Here U_{avg} denotes the average velocity.

A closed form of the turbulent time-averaged velocity profile cannot be derived for flow through a circular pipe. The measured turbulent profile shown in Fig. 3.5 has sharp gradients at the wall compared to that for a laminar flow. It is almost like a 'top hat' profile and it gives the impression that there is a slip at the wall. We may make the following assumptions for turbulent flow through the pipe: (a) Steady in time mean and axisymmetric and (b) fully developed. If we further assume that the law of the wall for a turbulent flow (Eq. 2.5) is valid all the way from the wall to the axis, we can directly deduce a relation between the friction factor and Reynolds number.

$$\frac{1}{f^{1/2}} = 1.99 \log_{10} Re_D f^{1/2} - 1.02 \tag{3.16}$$

Surprisingly, in spite of major simplifications this relation provides a fairly good agreement with the experimental data. The constants (1.99 and 1.02) in Eq. 3.16 can be adjusted to provide a better agreement with the experimental data.

In the same way parabolic velocity profile for the fully developed laminar flow between parallel plates can be obtained from the governing equations

$$u = u_{\mathrm{max}}(1 - y^2/h^2) \tag{3.17}$$

where h denotes half of the separation between parallel plates, u_{max} the centerline velocity, y the normal distance from the symmetry plane located at the middle of two plates. The difference between laminar and turbulent velocity profiles in this case is similar to that for circular pipe. For a turbulent flow between parallel plates, we can use the same assumptions as that for flow through a circular pipe and arrive at a relation between the friction factor and Reynolds number.

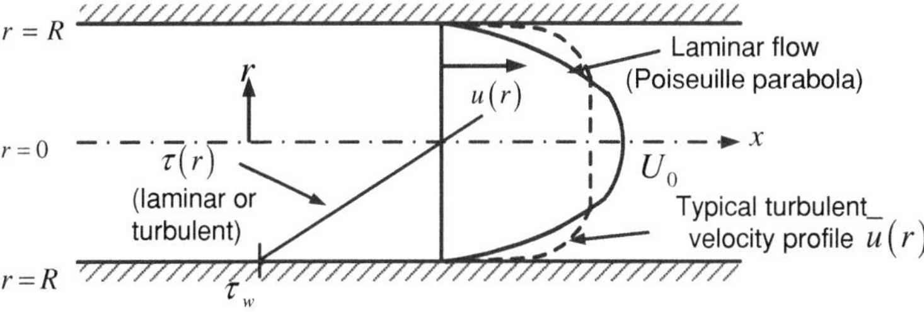

Fig. 3.5 A comparison of laminar and time-averaged turbulent fully developed velocity profiles through a circular pipe

Moody's diagram (Moody 1944) (Fig. 3.6) provides a composite plot of friction factor for a wide range of Reynolds number and non-dimensional wall roughness and covers both laminar and turbulent flow regimes. Darcy friction factor f is a strong function of the surface roughness in the turbulent flow regime. This is because even a small amount of roughness can disturb extremely thin wall layers. For a turbulent flow through a smooth pipe the friction factor decreases with an increase in Reynolds number. However for a rough pipe at a particular Reynolds number the value of f is higher than that for a smooth pipe. Further beyond a particular non-dimensional value of the wall roughness the friction factor in a turbulent flow depends only on the value of wall roughness and not on the value of Reynolds number, unlike that for a smooth pipe. A better non-dimensional way to represent the effect of roughness is by checking the value of the non-dimensional parameter $k/D\ Re_D$, where k denotes the roughness height. If $k/D\ Re_D < 10$ the effect of the roughness can be negligible and for $k/D\ Re_D > 1,000$ the flow can be assumed to be fully rough and therefore entirely independent of the value of Reynolds number (Fig. 3.6).

3.5 Separated Flows

Separated flows are widely encountered in engineering applications. A pressure gradient has a strong effect on the flow behavior within a boundary-layer. It can be shown that negative pressure gradient ($dp/dx < 0$) in the flow direction in a

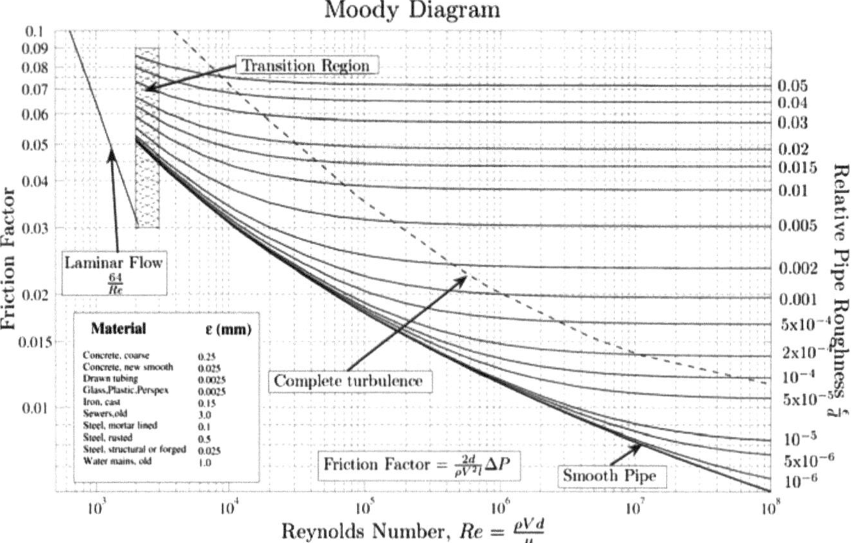

Fig. 3.6 Moody's diagram to calculate friction factor at different values of Reynolds number and wall roughness

boundary-layer, termed as the favourable pressure gradient, cannot lead to flow separation as it adds the flow or in other words competes with the retarding effects of the viscosity. In the same way, an increasing pressure in the flow direction ($dp/dx > 0$), termed as the adverse pressure gradient, resists the flow. This is because a moderate value of the adverse pressure gradient results in a point of inflection occurring in the velocity profile. A profile with a favourable pressure gradient has no point of inflection. The flow over the curved surface experiences favourable pressure gradient up to point B (Fig. 3.7). Further downstream the flow encounters adverse pressure gradient which adds to the retarding effect of the viscosity resulting in the thickening of the boundary layer. A continuous retardation of the flow results in a zero shear stress at the wall. A further adverse pressure gradient causes the flow to separate or in other words the flow does not follow the body contour. This is the reason that a diffuser with a large angle may lead to flow separation and thus poor performance. The location where the shear stress is zero is termed as the point of separation. In certain situations the separating flow may reattach to the body. The modeling of separated flows is a challenging test case for turbulence modelers. The simplified boundary-layer equations are not valid in the separated region that involves back flow and a strong interaction between the inner viscous region and the outer inviscid region. Therefore the Navier–Stokes equations need to be used to obtain the flow field. Compared to turbulent boundary layer, laminar boundary layers cannot resist adverse pressure gradient and therefore separates easily. However, turbulent boundary-layers involve good mixing but at the cost of large wall friction. Figure 3.8 shows two examples of configurations that cause two-dimensional flow separation and reattachment.

A classic example of separated flow is the one past a circular cylinder. This is perhaps one of the most widely investigated flow configuration. The flow approaching the cylinder is steady. However, the flow downstream may be highly unsteady and this behaviour downstream depends strongly on the Reynolds number of the approaching flow. The location of the point of separation of the flow from the cylinder is not fixed and it also depends on the Reynolds number of the approaching flow. We will see that a fairly accurate investigation of such flows can be obtained by using either LES or DNS (these methods of studying turbulence are discussed in Chap. 8). However, a treatment based on Reynolds-averaged Navier–Stokes equations (discussed in Chap. 4) may not show accurate results. Flows over

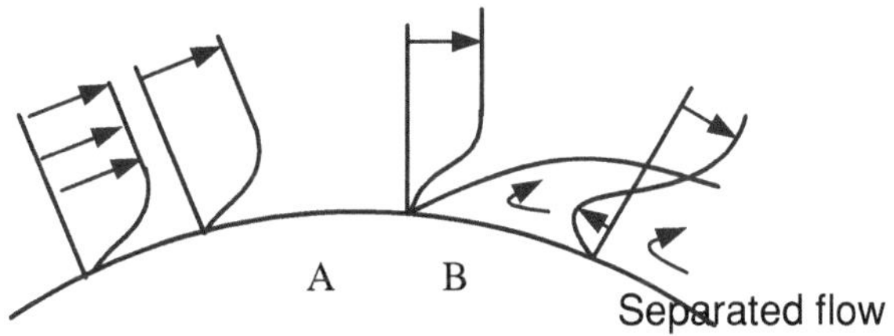

Fig. 3.7 Flow separation over a curved surface

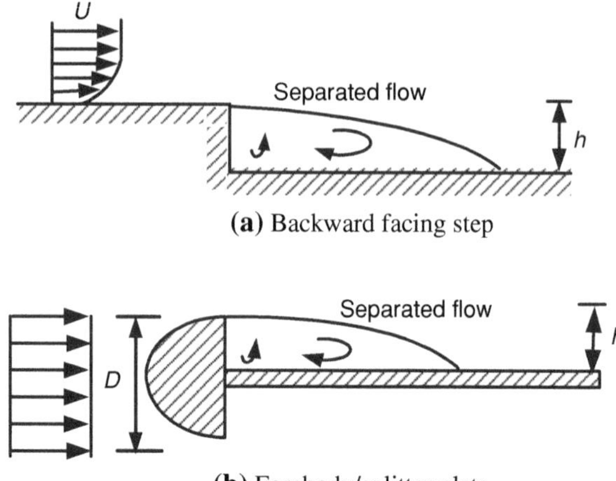

Fig. 3.8 Two common flow geometries causing flow separation

bodies with sharp corners can also cause separation such as the flow past a prism mounted on a flat plate.

3.6 Concluding Remarks

A presentation of detailed physics of important turbulent free-shear and wall-bounded flows in this chapter will enable a reader to appreciate the need for tackling turbulent flows in engineering. In the subsequent chapters we will present ways that can be used to model or simulate turbulent flows. These can be broadly divided in two categories: (a) ones employing time-averaged governing equations and (b) ones treating the instantaneous, three-dimensional turbulent flows. We will see that both approaches have inherent limitations: (a) first one has the closure problem associated with it, which requires a decision on choosing an appropriate turbulence model and (b) the second one has the limitation of either choosing a suitable subgrid model (in LES) or dealing with limited computer resources. We will deal with both broad approaches in detail in the subsequent chapters.

References

Blasius H (1908) Grenzschichten in Flussigkeiten mit kleiner Reibung. Z Math Phys 56:1–37
Incropera FP, DeWitt DP (2005) Fundamentals of heat and mass transfer, 5th edn. Wiley, USA
Klebanoff PS (1955) Characteristics of turbulence in a boundary-layer with zero pressure gradient. NACA report 1247
Liepmann HW (1943) Investigations on laminar boundary-layer stability and transition on curved boundaries. NACA wartime report w107 (ACR3H30)
Moody LF (1944) Friction factors for pipe flow. Trans ASME 66:671–684
White FM (2006) Viscous fluid flow, 3rd edn. McGraw-Hill, USA

Chapter 4
Reynolds-Averaged Governing Equations and Closure Problem

Abstract This chapter presents a widely used approach to model turbulent flows, i.e., to consider the time-averaged governing equations. It is shown that though this approach leads to a considerable simplification, an undesirable outcome is the closure problem, i.e., the number of unknowns exceeding the number of equations available. The closure problem sets the ground for a wide range and variety of RANS equations based turbulence models available in the literature.

4.1 Types of Reynolds-Averaging

All flow variables fluctuate in a turbulent flow so that they are always functions of spatial coordinates (x, y, z) and time (t) as

$$
\begin{aligned}
u &= u(x, y, z, t) \\
v &= v(x, y, z, t) \\
w &= w(x, y, z, t) \\
p &= p(x, y, z, t)
\end{aligned}
\tag{4.1}
$$

Here u, v and w denote the velocity components in three co-ordinate directions, respectively and p the pressure.

Using Reynolds (1895) decomposition one can express each turbulent flow variable as a sum of its mean and fluctuating components. This assumption means that for a general three-dimensional flow

$$
\begin{aligned}
u(x, y, z, t) &= \bar{u}(x, y, z) + u'(x, y, z, t) \\
v(x, y, z, t) &= \bar{v}(x, y, z) + v'(x, y, z, t) \\
w(x, y, z, t) &= \bar{w}(x, y, z) + w'(x, y, z, t) \\
p(x, y, z, t) &= \bar{p}(x, y, z) + p'(x, y, z, t)
\end{aligned}
\tag{4.2}
$$

A. Dewan, *Tackling Turbulent Flows in Engineering*,
DOI: 10.1007/978-3-642-14767-8_4, © Springer-Verlag Berlin Heidelberg 2011

Here bar (-) over a flow variable denotes its time-averaged component and symbol ($'$) its fluctuating component. In the subsequent sections we present three types of Reynolds-averaging that can be considered depending on a particular situation.

4.1.1 Time-Average

Engineers are primarily interested in time-averaged flow properties. The time average (or a mean) of a variable can be defined as a long time average.

$$\bar{u}(x,y,z) = \lim_{T \to \infty} \frac{1}{T} \int\limits_{to}^{to+T} u(x,y,z,t)\mathrm{d}t \tag{4.3}$$

The averaging time T should be a sufficiently long compared to the typical oscillations of the concerned flow property. This type of averaging is appropriate for stationary turbulence which is the one whose time average does not vary with time.

It is important to note that the fluctuating components are usually three-dimensional in a two-dimensional time-averaged flow. For example, time-averaged boundary layer flow past a flat plate may be steady and two-dimensional, but an instantaneous flow (i.e., without any averaging) may be highly three-dimensional and unsteady.

4.1.2 Spatial-Average

In the spatial average the flow properties are averaged over the space and therefore the averaged quantity remains unsteady.

$$\bar{u}(t) = \lim_{V \to \infty} \frac{1}{V} \iiint u(x,y,z,t)\mathrm{d}V \tag{4.4}$$

Here V denotes the averaging volume. This type of averaging is appropriate for homogeneous turbulence, i.e., the one whose space average is uniform in all flow directions.

4.1.3 Ensemble-Average

Ensemble average is the most general average and is also valid for time dependent mean flows. Here neither spatial nor temporal variation is removed from the averaged quantity. This essentially means that results (i.e., flow properties) from a large number of identical experiments are averaged by using the following expression

$$\bar{u}(x,y,z,t) = \lim_{N\to\infty} \frac{1}{N} \sum_{i=1}^{N} u(x,y,z,t) \tag{4.5}$$

where N denotes large number of identical experiments.

For a turbulent flow that is both stationary and homogeneous, the three averages are equal. Sometimes a turbulent flow may involve a slow time variation of the mean flow compared to quick transients inherent in all turbulent flows, for example, oscillating flow through a pipe. We may need to time-average such flow. In such cases, a time-averaged flow property can have small temporal variation provided inherent transients are very large (Fig. 4.1).

$$\phi(x,y,z,t) = \bar{\phi}(x,y,z,t) + \phi'(x,y,z,t)$$

$$\bar{\phi}(x,y,z,t) = \frac{1}{T} \int_{t+T_1}^{t+T_2} \phi(x,y,z,t)\mathrm{d}t \quad T_2 \le T \le T_1 \tag{4.6}$$

Equation 4.6 is valid for unsteady mean turbulent flow if T_1 and T_2 are significantly different. If T_1 and T_2 are of the same magnitudes then Eq. 4.6 cannot be used. In such a situation treatment of instantaneous flow using either direct numerical simulation or large eddy simulation (described in Chap. 8) may be appropriate.

Some common characteristics of time-average of two flow properties g and h are

$$\overline{g'} = 0, \quad \overline{\bar{g}h} = \bar{g}\bar{h}, \quad \overline{\bar{\bar{g}}} = \bar{g}$$

$$\overline{g+h} = \bar{g} + \bar{h}, \quad \overline{gh} = \bar{g}\bar{h} + \overline{g'h'}, \quad \overline{g'\bar{h}} = 0 \tag{4.7}$$

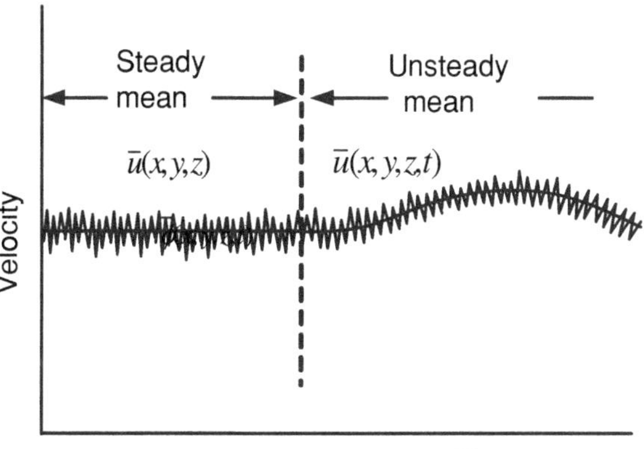

Fig. 4.1 Steady and unsteady mean turbulent flows

Time average commutes with the spatial differentiation and integration

$$\overline{\int g \mathrm{d}y} = \int \bar{g} \mathrm{d}y, \quad \overline{\frac{\partial g}{\partial k}} = \frac{\partial \bar{g}}{\partial k} \tag{4.8}$$

Further time averaging is a linear operation

$$\overline{a_1 g + b_1 h} = a_1 \bar{g} + b_1 \bar{h} \tag{4.9}$$

Here a_1 and b_1 denote two constants.

4.2 RANS and Scalar Equations

The Reynolds-averaged Navier–Stokes (RANS) equations are obtained by substituting each variable as a sum of the mean and time-averaged component as in Eq. 4.2 into the instantaneous Navier–Stokes equations and by time averaging the resulting equations. New terms result in the RANS equations from the non-linear transport terms in the momentum equations.

Consider the instantaneous continuity and Navier–Stokes equations for an incompressible flow with a constant viscosity:

$$\mathrm{div}\,\mathbf{u} = 0 \tag{4.10}$$

$$\frac{\partial u}{\partial t} + \mathrm{div}(u\mathbf{u}) = -\frac{1}{\rho}\frac{\partial p}{\partial x} + \frac{1}{\rho}\mathrm{div}(\mu\,\mathrm{grad}\,u) \tag{4.11}$$

$$\frac{\partial v}{\partial t} + \mathrm{div}(v\mathbf{u}) = -\frac{1}{\rho}\frac{\partial p}{\partial y} + \frac{1}{\rho}\mathrm{div}(\mu\,\mathrm{grad}\,v) \tag{4.12}$$

$$\frac{\partial w}{\partial t} + \mathrm{div}(w\mathbf{u}) = -\frac{1}{\rho}\frac{\partial p}{\partial z} + \frac{1}{\rho}\mathrm{div}(\mu\,\mathrm{grad}\,w) \tag{4.13}$$

The transport equation for a scalar variable (ϕ) may be written as

$$\frac{\partial \phi}{\partial t} + \mathrm{div}(\phi\mathbf{u}) = \mathrm{div}(\Gamma_\phi\,\mathrm{grad}\,\phi) + S_\phi \tag{4.14}$$

Here $\mathbf{u}$ denotes the velocity vector and u, v and w its components in three coordinate directions, respectively.

Replace each flow property by a sum of its mean and fluctuating component, and time average the resulting Eqs. 4.10–4.14 to obtain

$$\overline{\mathrm{div}\,\mathbf{u}} = \mathrm{div}\,\bar{\mathbf{u}} = 0 \tag{4.15}$$

$$\frac{\partial \bar{u}}{\partial t} + \mathrm{div}(\bar{u}\bar{\mathbf{u}}) + \mathrm{div}(\overline{u'\mathbf{u}'}) = -\frac{1}{\rho}\frac{\partial P}{\partial x} + \frac{1}{\rho}\mathrm{div}(\mu\,\mathrm{grad}\,\bar{u}) \tag{4.16}$$

The process of performing time average introduces one new term on the LHS of the resulting time-averaged momentum equation (4.16). The additional vector term $(\overline{u'\mathbf{u}'})$ has three components in three co-ordinate directions. Equation 4.16 is termed as the Reynolds-averaged Navier–Stokes equations and may be written for each co-ordinate direction as

$$\frac{\partial \bar{u}}{\partial t} + \text{div}(\bar{u}\bar{\mathbf{u}}) = -\frac{1}{\rho}\frac{\partial P}{\partial x} + \frac{1}{\rho}\text{div}(\mu \text{ grad } \bar{u}) + \left[-\frac{\partial \overline{u'u'}}{\partial x} - \frac{\partial \overline{u'v'}}{\partial y} - \frac{\partial \overline{u'w'}}{\partial z} \right] \quad (4.17)$$

$$\frac{\partial \bar{v}}{\partial t} + \text{div}(\bar{v}\bar{\mathbf{u}}) = -\frac{1}{\rho}\frac{\partial P}{\partial y} + \frac{1}{\rho}\text{div}(\mu \text{ grad } \bar{v}) + \left[-\frac{\partial \overline{v'u'}}{\partial x} - \frac{\partial \overline{v'v'}}{\partial y} - \frac{\partial \overline{v'w'}}{\partial z} \right] \quad (4.18)$$

$$\frac{\partial \bar{w}}{\partial t} + \text{div}(\bar{w}\bar{\mathbf{u}}) = -\frac{1}{\rho}\frac{\partial P}{\partial z} + \frac{1}{\rho}\text{div}(\mu \text{ grad } \bar{w}) + \left[-\frac{\partial \overline{w'u'}}{\partial x} - \frac{\partial \overline{w'v'}}{\partial y} - \frac{\partial \overline{w'w'}}{\partial z} \right] \quad (4.19)$$

The corresponding time-averaged transport equation for a scalar flow property ϕ is given as

$$\frac{\partial \Phi}{\partial t} + \text{div}(\Phi \mathbf{u}) = \text{div}(\Gamma_\Phi \text{grad } \Phi) + S_\Phi + \left[-\frac{\partial \overline{u'\varphi'}}{\partial x} - \frac{\partial \overline{v'\varphi'}}{\partial y} - \frac{\partial \overline{w'\varphi'}}{\partial z} \right] \quad (4.20)$$

Here Γ_ϕ denotes molecular diffusion of scalar flow variable and S_ϕ the source term in the scalar transport equation. The additional turbulent transport terms also arise in the time-averaged transport equation (4.20) for the scalar variable ϕ and represent transport of ϕ in x-, y- and z-directions, respectively, due to turbulent eddies.

4.3 Closure Problem

The instantaneous continuity and three Navier–Stokes equations form a closed set of four equations with four unknowns, namely, u, v, w and p. However, the Reynolds-averaged Navier–Stokes equations contain six additional terms denoted as the Reynolds stresses (actually nine Reynolds stress components reduce to six components due to the symmetry of this tensor τ_{ij}). In addition turbulent transport terms appear in the transport equations for scalar quantities. The additional terms are

Three turbulent normal stresses: $\tau_{xx} = -\rho\overline{u'^2}$, $\tau_{yy} = -\rho\overline{v'^2}$, $\tau_{zz} = -\rho\overline{w'^2}$.

Three turbulent shear stresses: $\tau_{xy} = \tau_{yx} = -\rho\overline{u'v'}$, $\tau_{xz} = \tau_{zx} = -\rho\overline{u'w'}$, $\tau_{yz} = \tau_{zy} = -\rho\overline{v'w'}$.

Three turbulent scalar fluxes: $\overline{u'\varphi'}$, $\overline{v'\varphi'}$ and $\overline{w'\varphi'}$.

Turbulent normal stresses denote the turbulent transport of momentum in the respective directions and turbulent shear stresses denote the turbulent transport of momentum from one direction to the one in its perpendicular direction, for example, from x-direction to y-direction in τ_{xy}. All these terms lead to the transport of momentum and scalar quantity and hence lead to better mixing compared to that in a laminar flow, where it is due to the molecular effects only.

It is clear that the number of unknowns is more than the number of equations available. However, in a laminar flow the number of unknowns are four (u, v, w and p) and the number of equations available are four (three momentum equations and one continuity equation). Thus we see that the system of governing equations in a laminar flow is closed, while in a turbulent flow it is not closed. It is the main task of turbulence modelling to develop mathematical models of sufficient accuracy and generality for engineers to predict the Reynolds stresses and the scalar transport terms and consequently mean flow and scalar fields.

Several types of turbulence models can be used to close the RANS equations, which include the simplest zero-equation model to the most complex Reynolds-stress transport models. Of these, the two equation models of turbulence have been widely applied to study engineering flows. We will see details of these models in Chaps. 5 and 6.

4.4 Concluding Remarks

In this chapter, we have considered different types of averaging of flow properties that can be used to simplify the treatment of turbulent flows. We have seen that a major simplicity results by using the Reynolds-averaging, i.e., the dependence of flow properties on spatial and temporal co-ordinates is reduced. However, it also leads to additional complexity, i.e., closure problem. In the next chapter we will look into some of the ways that can be adopted to address the closure problem. See Hanjalic (2005) for a discussion on the advantages and disadvantages of RANS approach vis-à-vis LES.

References

Hanjalic K (2005) Will RANS survive LES? A view of perspectives. Trans ASME J Fluids Eng 127:831–839
Reynolds O (1895) On the dynamical theory of incompressible viscous fluids and the determination of criterion. Phil Trans Royal Soc Lond Ser A 186:123

Chapter 5
Models Based on Boussinesq Approximation

Abstract This chapter presents the simplest way to handle the Reynolds-averaged Navier–Stokes (RANS) equations by using the Boussinesq approximation to close the system of equations. We also present a wide variety of turbulence models that can be employed using the Boussinesq approximation. It is shown that though this assumption involves major simplification, it leads to fairly accurate predictions in many simple situations. Some such situations are illustrated in this chapter.

5.1 Boussinesq Approximation

According to the Boussinesq approximation, the Reynolds stresses (τ_{ij}) can be expressed in terms of the mean strain rate (or turbulent momentum transport is assumed proportional to the mean strain rate).

$$\tau_{ij} = \mu_t \left(\frac{\partial \bar{u}_i}{\partial x_j} + \frac{\partial \bar{u}_j}{\partial x_i} \right) \tag{5.1}$$

Here μ_t is the constant of proportionality and denotes the turbulent or eddy viscosity. Since, we are primarily interested in the mean flow field, which in turn depends on Reynolds stresses, the key issue here is the computation of the eddy viscosity μ_t by using a suitable prescription. The eddy viscosity is a flow property and not a fluid property (unlike molecular viscosity) and therefore depends on flow characteristics. For example, consider two flow situations with air as the working fluid: (a) incompressible steady flow through a circular duct and (b) incompressible unsteady flow through a piston–cylinder arrangement. In both flow configurations, characteristics of flow and therefore the eddy viscosities may be different though it is the same working fluid (air) with the same value of the molecular viscosity.

A. Dewan, *Tackling Turbulent Flows in Engineering*,
DOI: 10.1007/978-3-642-14767-8_5, © Springer-Verlag Berlin Heidelberg 2011

The turbulent transport of heat, mass or other scalar quantities $(\overline{u_i'\varphi'})$ is modeled similar to that for momentum (taken to be proportional to the gradient of mean value of the transported quantity), i.e.

$$-\rho\overline{u_i'\varphi'} = \Gamma_t \left(\frac{\partial \bar{\phi}}{\partial x_i}\right) \tag{5.2}$$

where Γ_t the constant of proportionality denotes the turbulent diffusivity of a scalar variable. Turbulent transport of momentum or mass or heat occurs due to the movement of eddies. Eddies physically move a particular flow property from one location to other and also disperse (diffuse) the property while moving it from one location to another. Both these processes result in mixing of a particular flow property.

Since turbulent transport of momentum or heat or mass is due to the same mechanism, i.e. due to eddy mixing, we define a turbulent Prandtl number (Pr_t) for heat transfer and turbulent Schmidt number (Sc_t) for mass transfer as

$$Pr_t = \frac{\mu_t}{\Gamma_t} \text{ and } Sc_t = \frac{\mu_t}{m_t}. \tag{5.3}$$

Here m_t denotes the turbulent diffusivity of mass. A typical value of turbulent Prandtl number or turbulent Schmidt number used in engineering computations used is approximately 1.0.

5.2 Models Based on Boussinesq Approximation

Treatment of turbulence gets considerably simplified by the Boussinesq approximation because Reynolds stresses (τ_{ij}) and turbulent transport quantities $(\overline{u_i'\varphi'})$ are related to the mean flow $(\bar{u}, \bar{v}$ and $\bar{w})$ and scalar $(\bar{\phi})$ fields, respectively. Models based on the Boussinesq approximation can be characterized based on the number of transport equations used in addition to the momentum equations to compute the eddy viscosity. These models are basically of three types: (a) zero equation (or mixing length) models; (b) one equation models; and (c) two equation models. We will see that of these, the first two require the specification of at least one flow variable for a particular flow configuration and therefore are incomplete models. Two equation models are the simplest complete models and therefore are widely used. In the subsequent sections, we will consider these models in detail.

5.2.1 Mixing Length Models

In mixing length models, we assume that the eddy viscosity can be expressed as a product of a turbulent velocity scale and a length scale. The velocity scale is assumed to be related to the mean flow properties and length scale is assumed to be related to some typical width of the flow.

The Prandtl's (1925) mixing length model is one such model and is given for a thin-shear layer (such as, boundary-layer, jet and plume) as

$$v_t = l_{\text{mix}}^2 \left| \frac{\partial \bar{u}}{\partial y} \right| \tag{5.4}$$

where l_{mix} denotes the mixing length. The basic idea behind the mixing length hypothesis is that turbulent eddies (defined as the bunch of fluid particles with similar flow characteristics) while moving around in the fluid typically retain their momentum in x-direction over a distance in the y-direction equal to the mixing length (l_{mix}). Like molecular viscosity, eddy viscosity (μ_t) can also be related to the corresponding kinematic eddy viscosity (v_t) by a factor of density (ρ).

$$v_t = \frac{\mu_t}{\rho}. \tag{5.5}$$

Mixing length models need modifications to damp the eddy viscosity to zero as flow approaches a solid wall and also to account for the outer intermittent region. Because of their simplicity these were the earliest models used for the computation of turbulent flows. The mixing length model can also be used to predict turbulent diffusivity for the transport of a scalar variable by using the value of turbulent Prandtl number or turbulent Schmidt number approximately equal to one.

The first question that arises is, what is the value of the mixing length? Unfortunately, no universal prescription can be provided for different flow configurations. For free shear flows the mixing length is taken to be proportional to the width of the flow ($l_{\text{mix}} = cl$, where c denotes the coefficient of proportionality and l the width of the flow). Table 5.1 provides some typical constant of proportionality that are used for free-shear flows. It is clear that as a free shear flow expands in the streamwise direction, its mixing length also grows in proportion to its width in a free shear flow. Note that the constant of proportionality (c) is different in four cases.

For the wall bounded flows in the inner region, the mixing length is computed as $l_{\text{mix}} = ky$, where k denotes the von Karman constant and y the distance from the wall. A damping function is included to damp the eddy viscosity as the flow approaches a wall and reduce it to zero at the wall. Since no additional transport equation is used, such models are called the zero equation or algebraic model.

Table 5.1 Constant of proportionality for different turbulent free-shear flows

Flow	Constant of proportionality
Plane jet	0.09
Round jet	0.075
Far wake	0.16
Mixing layer	0.07

Mixing length models developed by Baldwin and Lomax (1978) and Cebeci and Smith (1974) are the most widely used turbulent models for the aerodynamics calculations. We present in some details the model of Cebeci and Smith (1974). It is two-layer model and the eddy viscosity is given as

$$\mu_t = \begin{cases} \mu_{ti} & \text{for } y \le y_m \\ \mu_{to} & \text{for } y > y_m \end{cases}. \tag{5.6}$$

The eddy viscosities are computed simultaneously by the two expressions given in Eq. 5.6 and at a particular location y_m where the outer eddy viscosity μ_{to} is more than the inner viscosity μ_{ti}, a switch over to the outer eddy viscosity takes place. The eddy viscosity in the inner layer is computed as

$$\mu_{ti} = \rho l_{mix}^2 \left[\left(\frac{\partial \bar{u}}{\partial y} \right)^2 + \left(\frac{\partial \bar{v}}{\partial x} \right)^2 \right]^{1/2}. \tag{5.7}$$

Equation 5.7 is a modified general form of Eq. 5.4 which was originally proposed for thin shear layers such as jet or boundary-layer. The mixing length in Eq. 5.7 is given as

$$l_{mix} = ky \left[1 - e^{-y^+/A_o^+} \right]. \tag{5.8}$$

The eddy viscosity in the outer layer is computed as

$$\mu_{to} = \alpha \rho U_e \delta_v^* F_{Kleb}. \tag{5.9}$$

U_e denotes the free stream velocity, F_{Kleb} is a curve fit by Klebanoff (1955) to the outer intermittency of turbulent flows. δ_v^* denotes the velocity thickness and for incompressible turbulent flows it reduces to the displacement thickness (Eq. 3.1). The model constants are $k = 0.41$, $\alpha = 0.0168$ and the variable A_0^+ is given as

$$A_0^+ = 26 \left[1 + y \frac{dP/dx}{\rho u_\tau^2} \right]^{-1/2}. \tag{5.10}$$

A_0^+ takes care of the dependence of the mixing length on the pressure gradient. The zero equations models have several deficiencies, which include

a. These do not directly account for the flow history effects, as the eddy viscosity is related to local mean flow properties.
b. Eddy viscosity reduces to zero when the mean strain rate equals zero, but this condition may not be valid in all cases.
c. These models cannot be directly applied to three-dimensional flows without any modification.
d. These are incomplete models because the mixing length needs to be specified. In addition, a prescription of the mixing length is not unique and depends on a particular flow configuration being studied.

e. The formulation of the model becomes difficult if there is a sudden change in the flow conditions. For example, the above mentioned prescription of the eddy viscosity for wall bounded flows can be used up to the trailing edge of aerofoil and subsequently the flow has the characteristics of separated flow and it is difficult to prescribe the mixing length for the separated region.

5.2.2 One Equation Model

In a one equation model, one transport equation in addition to the continuity and momentum equations is solved. The extra transport equation used can be for any turbulence variable. Use of an additional transport equation (with convection, diffusion and source terms) brings a turbulence model closer to reality. The most widely used one equation models are based on the transport equation for turbulence kinetic energy. In the subsequent sections, we will present some important one equation models.

Based on Equation for Turbulence Kinetic Energy

The instantaneous kinetic energy of a turbulent flow is the sum of the mean kinetic energy K and the turbulence kinetic energy k:

$$K + k = \left[\frac{1}{2} \left(\bar{u}^2 + \bar{v}^2 + \bar{w}^2 \right) \right] + \left[\frac{1}{2} \left(\overline{u'^2} + \overline{v'^2} + \overline{w'^2} \right) \right]. \qquad (5.11)$$

The exact transport equation for turbulence kinetic energy (k) may be obtained from the trace of transport equation for the Reynolds stress tensor τ_{ij}. The steps for the derivation of the exact transport equation for τ_{ij} are presented in Chap. 7. The equation is reproduced here:

$$\underbrace{\frac{\partial(\rho\overline{u_i'u_j'})}{\partial t}}_{} + \underbrace{\frac{\partial}{\partial x_k}\left(\rho\bar{u}_k\overline{u_i'u_j'}\right)}_{C_{ij}\,=\,\text{Convection}} = -\underbrace{\frac{\partial}{\partial x_k}\left[\rho\overline{u_i'u_j'u_k'} + \overline{p'(\delta_{kj}u_i' + \delta_{ik}'u_j')}\right]}_{D_{ij}^T\,=\,\text{Turbulent Diffusion}}$$

$$+ \underbrace{\frac{\partial}{\partial x_k}\left[\mu\frac{\partial}{\partial x_k}\left(\overline{u_i'u_j'}\right)\right]}_{D_{ij}^L\,=\,\text{Molecular Diffusion}} - \underbrace{\rho\left(\overline{u_i'u_k'}\frac{\partial\bar{u}_j}{\partial x_k} + \overline{u_j'u_k'}\frac{\partial\bar{u}_i}{\partial x_k}\right)}_{P_{ij}\,=\,\text{Stress Production}}$$

$$+ \underbrace{\overline{p'\left(\frac{\partial u_i'}{\partial x_j} + \frac{\partial u_j'}{\partial x_i}\right)}}_{\phi_{ij}\,=\,\text{Pressure Strain}} - \underbrace{2\mu\overline{\frac{\partial u_i'}{\partial x_k}\frac{\partial u_j'}{\partial x_k}}}_{\varepsilon_{ij}\,=\,\text{Dissipation}} \qquad (5.12)$$

Equation 5.12 is symmetric in tensor i and j. The exact transport equation for turbulence kinetic energy by substituting $i = j$ in Eq. 5.12 can be written as

$$\frac{\partial k}{\partial t} + \overline{u}_j \frac{\partial k}{\partial x_j} = \frac{\partial}{\partial x_j}\left(v\frac{\partial \overline{k}}{\partial x_j}\right) - v\overline{\frac{\partial u_i'}{\partial x_j}\frac{\partial u_i'}{\partial x_j}} - \overline{u_i'u_j'}\frac{\partial \overline{u}_i}{\partial x_j} - \frac{\partial}{\partial x_j}\left(\frac{1}{2}\overline{u_i'u_i'u_j'} + \frac{\overline{u_j'p'}}{\rho}\right) \qquad (5.13)$$

Note that the pressure strain term ϕ_{ij} in Eq. 5.12 reduces to zero by substituting $i = j$ in an incompressible flow.

Different terms in Eq. 5.13 from left to right can be interpreted as: (a) time rate of change of turbulence kinetic energy; (b) convection of turbulence kinetic energy by the mean flow; (c) molecular diffusion of turbulence kinetic energy; (d) dissipation of turbulence kinetic energy which denotes its conversion to thermal energy due to viscous effects; (e) production of turbulence kinetic energy by Reynolds stress acting on mean velocity gradient; and (f) transport of turbulence kinetic energy by fluctuating velocity and by pressure-velocity correlation. Equation 5.13 shows that there is a balance between all the terms. Of these, the second and last terms on the RHS of Eq. 5.13 require modeling. The remaining terms are exact and do not require any modeling.

To compute the Reynolds stress term using a one equation model, an extended Boussinesq relationship is used according to which the Reynolds stress tensor is modeled as

$$\tau_{ij} = -\rho\overline{u_i'u_j'} = \mu_t\left(\frac{\partial \overline{u}_i}{\partial x_j} + \frac{\partial \overline{u}_j}{\partial x_i}\right) - \frac{2}{3}\rho k\delta_{ij}. \qquad (5.14)$$

The second term on RHS which involves δ_{ij}, the Kronecker delta ($\delta_{ij} = 1$ if $i = j$ and $\delta_{ij} = 0$ if $i \neq j$) is added to make the expression applicable in the situation when $i = j$. This results in the expression for the Reynolds normal stress in an incompressible flow in a particular direction depending on the co-ordinate system used. For $i = j$

$$-\rho\overline{u_i'u_i'} = -\rho(\overline{u'^2} + \overline{v'^2} + \overline{w'^2}) = \mu_t\left(2\frac{\partial \overline{u}_i}{\partial x_i}\right) - \frac{2}{3}\rho k \qquad (5.15)$$

The first term on the RHS is zero due to continuity equation for an incompressible flow and therefore Eq. 5.15 is satisfied. The rate of dissipation of turbulence kinetic energy in Eq. 5.13 is modeled as

$$\varepsilon = C_{\mathrm{D}}k^{3/2}/l \qquad (5.16)$$

where C_{D} is a model constant and l a turbulent length scale that needs to be specified. Equation 5.16 is based on the dimensional arguments and with the assumption that the dissipation is a function of flow properties and independent of fluid properties.

The sum of the turbulent transport and pressure fluctuation term is modeled based on the gradient diffusion hypothesis as

$$\frac{\partial}{\partial x_j}\left(\frac{1}{2}\overline{u_i'u_i'u_j'} + \frac{\overline{u_j'p'}}{\rho}\right) = -\frac{v_t}{\sigma_k}\frac{\partial k}{\partial x_j} \qquad (5.17)$$

Here σ_k denotes the turbulent Prandtl number for turbulence kinetic energy. σ_k is also a flow property and not a fluid property. The diffusions of momentum and turbulence kinetic energy are both due to turbulent fluctuations and therefore their ratio is found to be approximately equal to one.

Thus, the final modeled transport equation for turbulence kinetic energy is given as

$$\rho\frac{\partial k}{\partial t} + \rho\bar{u}_j\frac{\partial k}{\partial x_j} = \frac{\partial}{\partial x_j}\left[\left(\mu + \frac{\mu_t}{\sigma_k}\right)\frac{\partial k}{\partial x_j}\right] + \tau_{ij}\frac{\partial \overline{u}_i}{\partial x_j} - \varepsilon\rho \tag{5.18}$$

It may be noted that theoretically the dissipation causes loss of turbulence kinetic energy at the smallest scales due to viscous effects. However, the dissipation modelled using Eq. 5.16 may not correctly represent this behavior because in this equation we have no variable dealing with small scales of turbulence. The eddy viscosity is modeled as

$$\mu_t = \rho k^{1/2} l \tag{5.19}$$

A major disadvantage of this model is that it is an incomplete model, since we need to specify the length scale l. Further no unique prescription of the length scale can be specified.

Based on Equation for Eddy Viscosity

The Spalart and Allmaras model (1992) is a simple one equation model. It directly solves a modeled transport equation for the kinematic eddy (turbulent) viscosity itself. In this model, a length scale related to the local shear layer thickness need not be calculated. The relevant transport equation for Spalart–Allmaras variable is

$$\frac{\partial \tilde{v}}{\partial t} + \bar{u}_j\frac{\partial \tilde{v}}{\partial x_j} = c_{b1}(1 - f_{t2})\tilde{S}\tilde{v} - \left[c_{w1}f_w - \frac{c_{b1}}{k^2}f_{t2}\right]\left(\frac{\tilde{v}}{d}\right)^2 + \frac{1}{\sigma}\left[\frac{\partial}{\partial x_j}(v + \tilde{v})\frac{\partial \tilde{v}}{\partial x_j}\right]$$
$$+ c_{b2}\frac{\partial \bar{v}}{\partial x_i}\frac{\partial \bar{v}}{\partial x_i} \tag{5.20}$$

The kinematic turbulent viscosity is given as product of Spalart–Allmaras variable and f_{v1}. Model constants: $c_{b1} = 0.1355$, $c_{b2} = 0.662$, $c_{v1} = 7.1$, $\sigma = 2/3$, $c_{w1} = \frac{c_{b1}}{\kappa^2} + \frac{1+c_{b2}}{\sigma}$, $c_{w2} = 0.3$, $c_{w3} = 2.0$, $\kappa = 0.41$ and the model relations:

$$f_{v1} = \frac{\chi^3}{\chi^3 + c_{v1}^3}, f_{v2} = 1 - \frac{\chi}{1 + \chi f_{v1}}, f_w = g\left(\frac{1 + c_{w3}^6}{g^6 + c_{w3}^6}\right)^{1/6},$$

$$\chi = \frac{\tilde{v}}{v}, g = r + c_{w2}(r^6 - r), r = \min\left[\frac{\tilde{v}}{\tilde{S}\kappa^2 d^2}, 10\right], v_t = \tilde{v}f_{v1},$$

$$\tilde{S} = S + \frac{\tilde{v}}{\kappa^2 d^2}f_{v2}, S = \sqrt{2\Omega_{ij}\Omega_{ij}}, f_{t2} = c_{t3}\exp(-c_{t4}\chi^2), c_{t3} = 1.2, c_{t4} = 0.5 \tag{5.21}$$

Ω_{ij} denotes the rotation tensor $\Omega_{ij} = 1/2\left(\partial \overline{u_i}/\partial x_j - \partial \overline{u_j}/\partial x_i\right)$ and d the distance from the surface. The Spalart and Allmaras model gives good results for boundary layers subjected to adverse pressure gradients. This model was designed specifically for aerospace applications involving wall-bounded flows. It can also be used for turbo-machinery applications. The model of Baldwin and Barth (1990) is another one equation based on the transport equation for the eddy viscosity.

5.2.3 Two Equation Models

At least two variables (for example, velocity and length scales) are needed to characterize turbulent flows completely. Therefore, two equation models are the simplest complete models because two additional transport equations are employed to independently describe two turbulence variables. Several two equation models have been proposed in the literature, which include the k-ε model (Launder and Spalding 1974) and k-ω model (Wilcox 1989). Of these, the standard k-ε model is one of the most widely used turbulence models. We will see in details some of the widely used two equation models in the next chapter.

5.3 Limitations of Boussinesq Approximation

The key assumption that the turbulent stresses are proportional to the mean strain rate may not hold true in many situations. Further we assume an isotropic eddy viscosity (i.e. which is same in all the directions) and this assumption may also fail in some situations. There are many situations where these assumptions are not likely to be valid, for example, flows with significant streamline curvature, flows with large mean strain rates, and flows in rotating fluids. Wilcox (2006) presents an interesting summary of such flows, where the Boussinesq assumption is likely to fail.

5.4 Examples

Zero equation models work well in simple flows (which do not separate and where thin shear layer assumption is valid) such as jets (Fig. 5.1), mixing layers, wakes, boundary layer flow, flow through pipe, flow between parallel plates, etc.

5.5 Concluding Remarks

We must remember that we have made a major simplification by using the governing equations for the time-averaged flow field rather than for the instantaneous

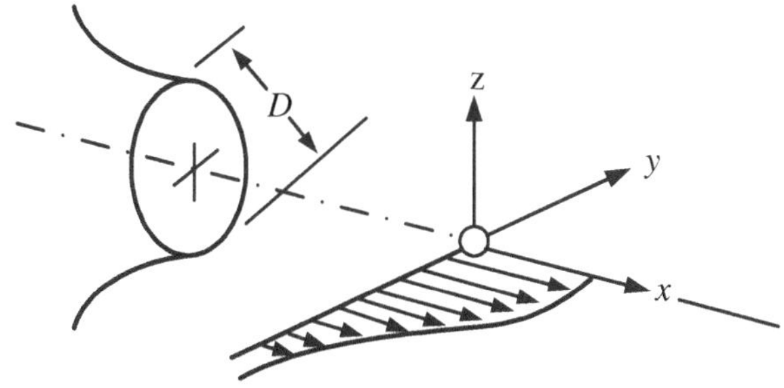

Fig. 5.1 Schematic of turbulent round jet

flow field. Thus, we cannot obtain the characteristics of instantaneous, three-dimensional flow field, energy spectrum, etc., from the numerical solution of the time-averaged governing equations. In this chapter, we have summarized different models that can be used to study turbulent flows based on the Boussinesq approximation. The two simplest models, namely, the zero equation and one equation models are incomplete models of turbulence because these need specification of at least one flow property. The zero equation models, based on the thin shear layer assumption, are known to work well in simple flows only. These were widely used in the initial stages of turbulence research. One equation models though physically more rigorous than the zero equation models have not enjoyed much acceptability because these models are almost half as complex as the simplest complete two equation models, but require specification of the length scale and are incomplete models and therefore not convenient to use. In the next chapter, we will present in details features of important two equation models.

References

Baldwin BS, Barth TJ (1990) A one-equation turbulence transport model for high Reynolds number wall-bounded flows. NASA TM-102847

Baldwin BS, Lomax H (1978) Thin-layer approximation and algebraic model for separated flows, AIAA paper 78–257, Huntsville, USA

Cebeci T, Smith AMO (1974) Analysis of turbulent boundary layers, Ser Appl Math Mech, vol 15, Academic Press, New York

Klebanoff PS (1955) Characteristics of turbulence in a boundary layer with zero pressure gradient, NACA TN 3178

Launder BE, Spalding DB (1974) The numerical computation of turbulent flow. Comput Methods Appl Mech Energy 3:269–289

Prandtl L (1925) Uber die ausgebildete Turbulenz. ZAMM 5:136–139

Spalart PR, Allmaras SR (1992) A one equation turbulence model for aerodynamic flows, AIAA paper 92–439, Reno, USA

Wilcox DC (1989) Reassessment of the scale determining equation for advanced turbulence models. AIAA J 26(11):1299–1310

Wilcox DC (2006) Turbulence modeling for CFD, 3rd edn. DCW Industries, California, pp 303–308

Chapter 6
k–ε and Other Two Equations Models

Abstract This chapter presents features of the standard k–ε model and some other two equation models. We first start with the steps for the derivations of the exact transport equations for k and ε and subsequently proceed to propose the modeled forms of these transport equations. We also suggest ways to handle walls and some other flow complexities using the k–ε model. We close this chapter by presenting the major potentials of the two equation models. We also present some examples where the k–ε model fails to accurately predict the flow behavior and needs to be modified.

6.1 Introduction

The two equation models are the simplest complete models of turbulence because two different transport equations are employed to characterize two independent turbulent flow properties. These models do not need specifications of any flow variable and therefore are convenient to use unlike zero- and one-equation models of turbulence.

6.2 Standard k–ε Model

The standard k–ε model is widely used in industrial turbulent flow and heat transfer computations mainly due to its robustness, computational economy, and reasonable accuracy for a wide variety of turbulent flows. It is somewhat a semi-empirical model, mainly because the modeled transport equation for dissipation used in the model depends on phenomenological considerations and empiricism. It is more expensive to implement than mixing length model or one equation model due to the use of two extra partial differential equations, as compared to zero- and

A. Dewan, *Tackling Turbulent Flows in Engineering*,   59
DOI: 10.1007/978-3-642-14767-8_6, © Springer-Verlag Berlin Heidelberg 2011

one-equation transport equation, respectively, employed in the former models. This model was originally proposed by Jones and Launder (1972).

6.3 Exact Transport Equations for k and ε

The steps for the derivation of the exact transport equation for turbulence kinetic energy were already presented in Sect. 5.2.2 in Chap. 5. The exact transport equation for the rate of dissipation of turbulence kinetic energy (ε) can be obtained from the following mathematical operation

$$\overline{2v\frac{\partial u_i'}{\partial x_j}\frac{\partial}{\partial x_j}[N(u_i)]} = 0 \tag{6.1}$$

where $N(u_i)$ denotes the Navier–Stokes operator and is defined as

$$N(u_i) = \rho\frac{\partial u_i}{\partial t} + \rho\, u_k\frac{\partial u_i}{\partial x_k} + \frac{\partial p}{\partial x_i} - \mu\frac{\partial^2 u_i}{\partial x_k \partial x_k} \tag{6.2}$$

The exact transport equation for the dissipation obtained after much algebra may be written as

$$\frac{\partial \varepsilon}{\partial t} + \bar{u}_j\frac{\partial \varepsilon}{\partial x_j} = -2v\left[\overline{u_{i,k}'u_{j,k}'} + \overline{u_{k,i}'u_{k,j}'}\right]\frac{\partial \bar{u}_i}{\partial x_j} - 2v\overline{u_k'u_{i,j}'}\frac{\partial^2 \bar{u}_i}{\partial x_k \partial x_j}$$

$$-2v\overline{u_{i,k}'u_{i,m}'u_{k,m}'} - 2v^2\overline{u_{i,km}'u_{i,km}'} + \frac{\partial}{\partial x_j}\left[v\frac{\partial \varepsilon}{\partial x_j} - \overline{vu_j'u_{i,m}'u_{i,m}'} - 2\frac{v}{\rho}\overline{p_{,m}'u_{j,m}'}\right] \tag{6.3}$$

Equation 6.3 is more complex compared to the exact equation for turbulent kinetic energy (5.12). Equation 6.3 has total ten terms. On LHS we have the standard unsteady and convection terms. On the RHS, however, we have several complex terms. These are denoted as the production of dissipation, dissipation of dissipation, turbulent transport and molecular diffusion of dissipation. Of these, the unsteady, convection and molecular diffusion terms do not require any modeling and the remaining seven terms require modeling. Unfortunately no assistance can be obtained from the experimental data to provide any guideline for modeling different terms of Eq. 6.3. As we will see in the next section, the modeled form of the dissipation equation used in the literature is a major weakness of the k–ε model.

6.4 Modelled Transport Equations for k and ε

We have already obtained the modeled form of the transport equation for turbulent kinetic energy in Sect. 5.2.2 in Chap. 5. In a two equation model a transport equation for dissipation of turbulence kinetic energy is solved whereas in a one equation model the dissipation is modeled using an algebraic relation (as seen in Sect. 5.2.2 in Chap. 5).

We have earlier presented the modeled transport equation for turbulence kinetic energy in the tensor notation (Eq. 5.18). The modeled transport equation for turbulent kinetic energy (k) can be written in the component form as

$$\frac{\partial k}{\partial t} + \bar{u}\frac{\partial k}{\partial x} + \bar{v}\frac{\partial k}{\partial y} + \bar{w}\frac{\partial k}{\partial z} = \frac{\partial}{\partial x}\left[\left(v + \frac{v_t}{\sigma_k}\right)\frac{\partial k}{\partial x}\right] + \frac{\partial}{\partial y}\left[\left(v + \frac{v_t}{\sigma_k}\right)\frac{\partial k}{\partial y}\right]$$
$$+ \frac{\partial}{\partial z}\left[\left(v + \frac{v_t}{\sigma_k}\right)\frac{\partial k}{\partial z}\right] + P_k - \varepsilon \tag{6.4}$$

The modeled transport equation for the dissipation in the tensor notation can be written as

$$\frac{\partial \varepsilon}{\partial t} + \bar{u}_j\frac{\partial \varepsilon}{\partial x_j} = \frac{\partial}{\partial x_j}\left[\left(v + \frac{v_t}{\sigma_\varepsilon}\right)\frac{\partial \varepsilon}{\partial x_j}\right] + C_{\varepsilon 1}\frac{\varepsilon}{k}\overline{u_i'u_j'}\frac{\partial \bar{u}_i}{\partial x_j} - C_{\varepsilon 2}\frac{\varepsilon^2}{k} \tag{6.5}$$

The net effect of the seven terms on the RHS of Eq. 6.3 is modelled by three terms on the RHS of Eq. 6.5.

The resulting transport equation for dissipation of turbulence kinetic energy (scalar) can be written in the expanded form as

$$\frac{\partial \varepsilon}{\partial t} + \bar{u}\frac{\partial \varepsilon}{\partial x} + \bar{v}\frac{\partial \varepsilon}{\partial y} + \bar{w}\frac{\partial \varepsilon}{\partial z} = \frac{\partial}{\partial x}\left[\left(v + \frac{v_t}{\sigma_\varepsilon}\right)\frac{\partial \varepsilon}{\partial x}\right] + \frac{\partial}{\partial y}\left[\left(v + \frac{v_t}{\sigma_\varepsilon}\right)\frac{\partial \varepsilon}{\partial y}\right]$$
$$+ \frac{\partial}{\partial z}\left[\left(v + \frac{v_t}{\sigma_k}\right)\frac{\partial \varepsilon}{\partial z}\right] + C_{\varepsilon 1}\frac{\varepsilon}{k}P_k - C_{\varepsilon 2}\frac{\varepsilon^2}{k} \tag{6.6}$$

where P_k denotes the rate of shear production of k and is given in the expanded form as

$$P_k = v_t\left[\left(\frac{\partial \bar{u}}{\partial y} + \frac{\partial \bar{v}}{\partial x}\right)^2 + \left(\frac{\partial \bar{v}}{\partial z} + \frac{\partial \bar{w}}{\partial y}\right)^2 + \left(\frac{\partial \bar{u}}{\partial z} + \frac{\partial \bar{w}}{\partial x}\right)^2\right]$$
$$+ v_t\left[2\left(\frac{\partial \bar{u}}{\partial x}\right)^2 + 2\left(\frac{\partial \bar{v}}{\partial y}\right)^2 + 2\left(\frac{\partial \bar{w}}{\partial z}\right)^2\right] \tag{6.7}$$

The eddy viscosity (kinematic) is given as $v_t = C_\mu\frac{k^2}{\varepsilon}$ and the following standard values of the model constants are used (Launder and Spalding 1974) $C_\mu = 0.09$, $\sigma_k = 1.00$, $\sigma_\varepsilon = 1.3$, $C_{\varepsilon 1} = 1.44$, $C_{\varepsilon 2} = 1.92$.

6.5 Features of the k–ε Model

1. It is a high turbulence Reynolds number ($Re_t = k^2/\varepsilon\nu$) model. Therefore it cannot be applied without suitable modifications in the regions with low Re_t, for example, in the vicinity of a solid wall, in laminar to turbulent transition, etc.
2. In this model the solution of two separate transport equations for k and ε allows the turbulent velocity and length scales to be independently determined. It is one of the simplest complete models of turbulence.
3. Each term of the modeled transport equation for k almost accurately represents the corresponding term in the exact equation. However, the gross effect of several terms in the exact dissipation equation is modeled by few terms, or in other words, there is no one to one correspondence between different terms in the modeled and exact transport equations for dissipation.

6.6 Boundary Conditions

We need to apply appropriate conditions at the boundaries for computing the flow in a particular computational domain. Here we present some commonly encountered boundary conditions using the standard k–ε model.

1. Inlet: Provide distributions of k and ε along with flow properties, i.e., velocity and temperature, in the corresponding real situation. In some cases, it is difficult to obtain values of k and ε at the inlet and in such cases these can be obtained based on an approximation from the turbulent intensity T_i and a characteristic length L of the flow configuration:

$$k = \frac{3}{2}\left(U_{ref}T_i\right)^2, \quad \varepsilon = C_\mu^{3/4}\frac{k^{3/2}}{l}, \quad l = 0.07L \tag{6.8}$$

Here l denotes a turbulent length scale.
2. Outlet: At the outlet usually turbulence (k and ε) is taken equal to zero, the mean temperature ($\overline{T_m}$) equal to the ambient temperature (T_∞) and pressure (p) equal to the atmospheric pressure p_∞.
3. Symmetry plane: Gradients of all flow properties normal to the plane or line of symmetry are taken equal to zero, i.e., $\partial\bar{u}_i/\partial n = 0, \partial\bar{T}/\partial n = 0, \partial k/\partial n = 0$ and $\partial\varepsilon/\partial n = 0$, where n denotes normal to the plane or line of symmetry.
4. The free-streams are usually non turbulent and therefore $k = 0$ and $\varepsilon = 0$ are usually specified. Mean velocity and temperature may be taken equal to their atmospheric counterparts.
5. At the solid wall either the no slip condition using the low-Re version or wall function approach can be applied. We will present the features of these two approaches in the next section.

6.7 Treatment of Wall

Turbulence in the presence of wall poses major complexities primarily because turbulence is zero at the wall and further turbulent flow field experiences very sharp gradients in the vicinity of a solid wall (Sect. 3.1 in Chap. 3). A turbulence model should be able to accurately capture these features. Two approaches can be adopted for the treatment of wall and we will see details of both these approaches in the subsequent sections.

6.7.1 Wall Functions Approach

One way to handle the near wall region is to employ the wall functions. The flow field near the wall is not solved, but obtained by semi-empirical functions termed as the wall functions. The first grid point is not located right at the wall but slightly away from the wall and the values of the flow variables at the first grid point obtained from the wall functions are used as the boundary conditions. The wall functions reduce the computational effort significantly because the region in the close proximity to the wall is not resolved. Several types of wall functions have been proposed in the literature. We will present details of some such functions. Figure 6.1 presents a schematic of these two approaches to treat a solid wall.

Standard Wall Functions

The standard wall functions for the mean streamwise velocity and turbulence kinetic energy are given as

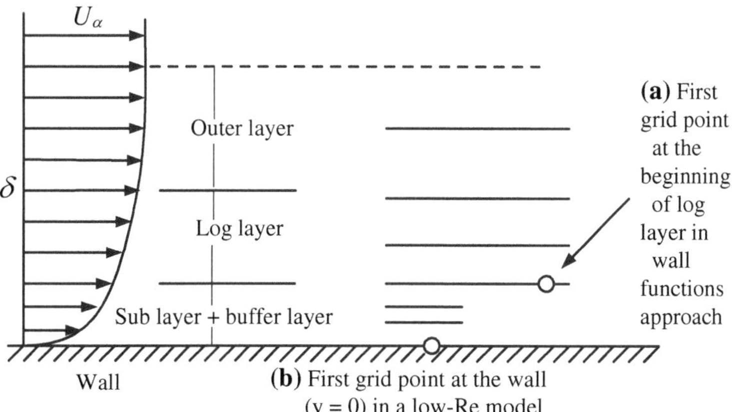

Fig. 6.1 **a** Wall function approach and **b** low Reynolds number version approach (dimensions not according to scale)

$$\overline{u}_p^+ = \frac{1}{\kappa} \ln y_p^+ + B \tag{6.9}$$

$$k_p = \frac{u_\tau^2}{C_\mu^{1/2}}, \; \varepsilon_p = \frac{u_\tau^3}{\kappa y_P} \tag{6.10}$$

where y_p^+ denotes the distance of the first grid point from the wall in the wall co-ordinates, $\kappa = 0.41$ the von Karman constant and $B = 5.0$ a dimensionless constant. The expression for velocity (Eq. 6.9) is based on the logarithmic law-of-the-wall which is assumed to be universally applicable in the vicinity of a solid wall irrespective of flow geometry and pressure gradient. The expressions for the kinetic energy and dissipation in Eq. 6.10 are the exact solution of the *k–ε* model equations in the wall region.

At high Reynolds number, the logarithmic near wall temperature distribution is used (Eq. 2.8)

$$T^+ = -\frac{(T_p - T_w)C_p \rho u_\tau}{q_w} = \mathrm{Pr}_t/k \, \ln y^+ + A(\mathrm{Pr}) \tag{6.11}$$

where T_p denotes the temperature at the near wall point y_p. T_w denotes the wall temperature, q_w the wall heat flux, C_p the fluid specific heat at constant pressure, Pr_t the turbulent Prandtl number and Pr the Prandtl number.

This approach has its limitations and therefore it is not universal. For example, in separating and reattaching flows and flows with strong curvature, the law of the wall is invalid and therefore the wall functions do not hold. In addition, the use of this approach involves some practical problems. A user needs to ensure that the location of the first grid point in the wall coordinates y_p^+ is in the logarithmic region. However, this cannot be determined a priori because the value of y_p^+ depends on the skin friction coefficient (C_f) whose value is obtained from the numerical solution. If y_p^+ is too close to the wall then the first grid point is located away from the region where the logarithmic law is valid and if it is farther away from the wall then the accuracy is reduced because the distance between the wall and first grid point becomes too large.

Non-Equilibrium Wall Functions

The non-equilibrium wall functions use the log-law for the mean velocity modified to account for the effects of the pressure gradient on the flow behaviour and use a two-layer approach to calculate the turbulence kinetic energy in the cells close to the wall. This improves its performance in complex flows. The modified logarithmic law applied is given as

$$\frac{\tilde{u}C_{\mu}^{1/4}k^{1/2}}{\tau_w/\rho} = \frac{1}{k}\ln\left(\frac{E\rho C_{\mu}^{1/4}k^{1/2}y}{\mu}\right) \tag{6.12}$$

$$\tilde{u} = \bar{u}\frac{1}{2}\frac{dp}{dx}\left[\frac{y_v}{\rho k\sqrt{k}}\ln\left(\frac{y}{y_v}\right) + \frac{y - y_v}{\rho k\sqrt{k}} + \frac{y_v^2}{\mu}\right] \tag{6.13}$$

here $E = 9.793$ and y_v denotes the viscous sub-layer thickness and is obtained from the expression

$$y_v = \frac{\mu y_v^*}{\rho C_{\mu}^{1/4}k_P^{1/2}} \tag{6.14}$$

where $y^*_v = 11.225$. Here two-layer concept is used to calculate the turbulence kinetic energy and dissipation rate at the first grid point. If $y < y_v$ the region is called the sublayer region and $y > y_v$ the region is called the fully turbulent region. Depending on the location of the first grid point with respect to y_v values of turbulence kinetic energy and dissipation are specified.

Enhanced Wall Treatments

Enhanced wall treatments employ two-layer model as in the non-equilibrium wall functions and enhanced wall functions with an objective to increase its applicability to different meshes. For fine meshes two-layer approach is used. The near-wall region is solved up to the viscous sublayer. The two layers are the viscosity-affected region where one-equation model of Wolfstein (1969) is used and the fully turbulent region where either the k–ε or RSM model is applied. The two-layer model smoothly merges with high-Re formulation of turbulent viscocity used in the turbulence models. For the coarse meshes enhanced wall-functions approach is used where a single wall law is applied throughout the wall region (i.e., viscous sublayer, buffer region, fully turbulent region) and linear (viscous) and logarithmic (turbulent) laws-of-the-wall are merged at the intermediate point.

Both non-equilibrium and enhanced wall functions are widely used in the commercial CFD softwares.

6.7.2 Low Reynolds Number Models

To accurately resolve the flow physics, the mesh needs to be extremely fine close to the wall and in low-Re models the flow field is solved all the way through the overlap layer and viscous sublayer right up to the wall ($y = 0$) with the use of the no slip condition at the wall. Thus this approach requires a very fine mesh.

Damping functions and source terms to the transport equations for k and ε need to be added to account for the low (turbulence) Reynolds number and wall effects. Several proposals exist in the literature. See for example, Patel et al. (1985) for an excellent review of eight important models up to 1985. New models continue to be developed. However, none of these models seem to be universally applicable. A comparison of different new low Re versions of the model is also an interesting research topic. The low-Re version of Launder and Sharma (1974) was among the first such models and we will present here details of their model.

In general, in a low-Re k–ε model the eddy viscosity is computed by a slightly modified expression

$$v_t = c_\mu f_\mu \frac{k^2}{\varepsilon} \tag{6.15}$$

Here f_μ denotes a damping function and in different models it is based either on the distance from the wall or turbulence Reynolds number. For example, in the Launder and Sharma model the damping function is given as

$$f_\mu = \exp\left[-\frac{2.5}{1 + Re_T/50}\right] \tag{6.16}$$

Far away from the wall this function becomes equal to 1.0 and thus the expression reduces to the standard expression for the eddy viscosity ($v_t = C_\mu k^2/\varepsilon$) used in the standard k–ε model. The following modified forms of the transport equations for k and ε are used in the Launder and Sharma (1974) model

$$\bar{u}\frac{\partial k}{\partial x} + v\frac{\partial k}{\partial y} = \frac{\partial}{\partial y}\left[\left(v + \frac{v_t}{\sigma_k}\right)\frac{\partial k}{\partial y}\right] + v_t\left(\frac{\partial k}{\partial y}\right)^2 - \tilde{\varepsilon} - 2v\left(\frac{\partial\sqrt{k}}{\partial y}\right)^2 \tag{6.17}$$

$$\bar{u}\frac{\partial\tilde{\varepsilon}}{\partial x} + v\frac{\partial\tilde{\varepsilon}}{\partial y} = \frac{\partial}{\partial y}\left[\left(v + \frac{v_t}{\sigma_k}\right)\frac{\partial\tilde{\varepsilon}}{\partial y}\right] + c_{\varepsilon 1}\frac{\tilde{\varepsilon}v_t}{k}\left(\frac{\partial\bar{u}}{\partial y}\right)^2 - c_{\varepsilon 2}\frac{\tilde{\varepsilon}^2}{k} + 2vv_t\left(\frac{\partial^2\bar{u}}{\partial y^2}\right)^2 \tag{6.18}$$

The modified form of the constant $c_{\varepsilon 2}$ used by Launder and Sharma (1974) is given as

$$c_{\varepsilon 2} = 1.92[1 - 0.3\exp(-Re_T^2)] \tag{6.19}$$

Note the additional one term each on the RHS of Eqs. 6.17 and 6.18 compared to that for the standard k–ε model. These additional terms vanish far away from the wall and thus this low-Re version of model reduces to the standard model far away from the wall.

Note that Launder and Sharma (1974) use a modified form of equation for the variable $\tilde{\varepsilon}$ which is similar to dissipation and one needs to specify $\tilde{\varepsilon} \approx 0$ at the wall. Dissipation does not vanish at a solid wall (it is in fact, quite large) and this behaviour of large dissipation at the wall is accounted by the additional term $2vv_t\left(\frac{\partial^2 u}{\partial y^2}\right)^2$ which assumes large values at the solid surface. The dissipation at the wall equal to zero can be used in conjunction with Eq. 6.18.

Another class of low Reynolds number turbulence model is based on the nonlinear (complex) constitutive relation between the Reynolds stress and mean flow characteristics. These models also do not use a constant value of c_μ as in the standard k–ε model but involve somewhat complex expression which returns a variable (Craft et al. 1996). These models are termed as nonlinear low-Re k–ε model. It is needless to say that these models contain a rather large number of constants and do not seem to be popular with industry.

In separated flows the Launder and Sharma k–ε model predicts high levels of near wall turbulence and therefore some modellers introduced some source terms to the dissipation equation and some modellers introduced another modification to the model by removing the dependence on the distance from the wall.

6.8 Example: Oscillatory Boundary Layers

A number of applications involve oscillatory boundary layers, for example, pulsating flow in arteries, flow past helicopter blades, etc. Turbulence models were originally proposed for steady flows and it is interesting to assess their behaviour for oscillatory flows. Sana et al. (2007) modified the damping functions of two low Reynolds number k–ε models originally based on the wall variables. The modifications were based on the Kolmogorov's length scale based on DNS data for steady boundary layers. The idea is that the wall shear stress becomes zero twice in an oscillatory flow and therefore its use to obtain the non-dimensional distance from the wall in the oscillatory flows is inappropriate. They modified the damping functions of the two low-Re versions of Myong and Kasagi (1990) and Nagano and Tagawa (1990). They compared their predictions with the DNS data of a wave boundary layer under sinusoidal pressure gradient for various wave Reynolds number and showed that the modified versions of the model perform better than the original models in predicting the near wall velocities and turbulence kinetic energy. They also observed that there is a scope for the improvement for the predictions during the adverse pressure gradient. They showed that the modified models also captured the interesting feature of a sudden increase in the wall shear stress during an adverse pressure gradient.

6.9 Some Modern Variants of k–ε Model

Some variants of the standard k–ε model have been proposed in the literature in order to enhance the range of applicability of this model. We present some important variants here.

6.9.1 RNG k–ε Model

RNG k–ε model was proposed by Yakhot and Orszag (1986) using a statistical technique called the renormalization group theory. It is similar in form to the standard k–ε model, but includes the following important refinements: (1) It has an additional term in its dissipation equation that is supposed to improve the accuracy for rapidly strained flows. (2) The effect of swirl on turbulence is included in the RNG model, thus enhancing its accuracy for swirling flows.

The RNG theory provides an analytical formula for turbulent Prandtl numbers, while the standard k–ε model uses the constant values. RNG theory also provides an analytically-derived differential formula for the effective viscosity that accounts for low-Reynolds-number effects. The transport equations for turbulent kinetic energy and its dissipation used in the RNG k–ε model are

$$\frac{\partial}{\partial t}(\rho k) + \frac{\partial}{\partial x_i}(\rho k \overline{u_i}) = \frac{\partial}{\partial x_j}\left[\left(\mu + \frac{\mu_t}{\sigma_k}\right)\frac{\partial k}{\partial x_j}\right] + P_k - \rho \varepsilon \tag{6.20}$$

$$\frac{\partial}{\partial t}(\rho \varepsilon) + \frac{\partial}{\partial x_i}(\rho \varepsilon \overline{u_i}) = \frac{\partial}{\partial x_j}\left[\left(\mu + \frac{\mu_t}{\sigma_\varepsilon}\right)\frac{\partial \varepsilon}{\partial x_j}\right] + c_{\varepsilon 1}\frac{\varepsilon}{k}P_k - c_{\varepsilon 2}^*\rho\frac{\varepsilon^2}{k} \tag{6.21}$$

The model constants and the auxiliary relations are $C_\mu = 0.0845, C_{\varepsilon 1} = 1.42,$ $C_{\varepsilon 2} = 1.68, \sigma_k = \sigma_\varepsilon = 0.7178, \beta = 0.012$

$$C_{\varepsilon 2}^* = C_{\varepsilon 2} + \frac{c_\mu \eta^3(1 - \eta/\eta_0)}{1 + \beta\eta^3} \quad \text{with} \quad \eta = \frac{Sk}{\varepsilon} \text{ and } \eta_0 = 4.38 \tag{6.22}$$

The values of the model constants are explicitly obtained in the RNG k–ε model.

6.9.2 Realizable k–ε Model

Realizable k–ε model was proposed by Shih et al. (1995) and differs from the standard k–ε model in two important ways: (1) It contains a new formulation for the turbulent viscosity and (2) it has a new transport equation for the dissipation rate ε which is derived from an exact equation for the transport of the mean-square vorticity fluctuation.

The following general expression for the normal Reynolds stress in an incompressible strained mean flow can be obtained by using the Boussinesq relationship and the standard expression for the eddy viscosity

$$\overline{u'^2} = \frac{2}{3}k - 2v_t\frac{\partial \overline{u}}{\partial x} \tag{6.23}$$

Therefore one obtains the result that the normal stress, $\overline{u'^2}$, which by definition is a positive quantity, becomes negative, i.e., "non-realizable", when the strain is large and therefore the following equation needs to be satisfied

$$\frac{k\partial \bar{u}}{\varepsilon \partial x} > \frac{1}{3C_\mu} \approx 3.7 \tag{6.24}$$

to ensure that the normal stress is positive. A possibility to ensure the realizability (positivity of normal stresses) is to make C_μ variable by sensitizing it to the mean flow (mean deformation) and the turbulence (k, ε).

As already mentioned a weakness of the standard k–ε model lies with the modeled equation for the dissipation rate (ε). The plane jet/round jet anomaly is considered to be mainly due to the modeled dissipation equation. The standard k–ε model correctly predicts the growth rate of the plane jet, but that of round jet is over predicted (Table 6.1) and this is termed as the plane jet/round jet anomaly.

The term "realizable" means that the model satisfies certain mathematical constraints on the Reynolds stresses. Both the standard k–ε model and the RNG k–ε model are not realizable. One limitation of the realizable k–ε model is that it produces non-physical turbulent viscosities in situations when computational domain contains both rotating and stationary fluid zones.

6.9.3 k–ω Model

The k–ω model originally given by Wilcox (1988) is based on the modelled transport equations for the turbulence kinetic energy (k) and the specific dissipation rate (ω), which can also be thought of as the ratio of ε to k. The model incorporates modifications for low-Reynolds-number effects and is applicable to wall-bounded flows without any further modifications and free shear flows. Transport equations for this model are given as

$$\frac{\partial k}{\partial t} + \bar{u}_j \frac{\partial k}{\partial x_j} = \frac{\partial}{\partial x_j}\left[\left(v + \sigma^* \frac{k}{\omega}\right)\frac{\partial k}{\partial x_j}\right] + \tau_{ij}\frac{\partial u_i}{\partial x_j} - \beta^* k\omega \tag{6.25}$$

$$\frac{\partial \omega}{\partial t} + \bar{u}_j \frac{\partial \omega}{\partial x_j} = \frac{\partial}{\partial x_j}\left[\left(v + \sigma \frac{k}{\omega}\right)\frac{\partial \omega}{\partial x_j}\right] + \alpha\frac{\omega}{k}\frac{\partial \bar{u}_i}{\partial x_j}\frac{\tau_{ij}}{\rho} - \beta\omega^2 \tag{6.26}$$

Table 6.1 Comparison of growth rates of four typical free shear flows predicted by k–ε and k–ω models (Wilcox 2006)

Flow	Measurements	k–ε model	k–ω model
Round jet	0.08–0.09	0.12	0.07–0.37
Plane jet	0.10–0.11	0.11	0.09–0.14
Far wake	0.36	0.25	0.30–0.50
Mixing layer	0.12	0.10	0.10–0.14

The model constants are given as

$$\alpha = \frac{5}{9}, \quad \beta = \frac{3}{40}, \quad \beta^* = \frac{9}{100}, \quad \sigma = \frac{1}{2}, \quad \sigma^* = \frac{1}{2}, \tag{6.27}$$

$$\varepsilon = \beta^* \omega k \tag{6.28}$$

This k–ω model is an improved version of the original model proposed by Kolmogorov (1942). The improvements mainly involve inclusion of the molecular diffusion and production terms to the original transport equation for the specific dissipation rate used by Kolmogorov (1942).

Impinging jets are relevant to many industrial applications and are quite complex due to an interaction of entrainment, mixing layers, stagnation lines and curvature of the streamlines in the impact region. Fernández et al. (2007) compared the performances of standard k–ε model, realizable k–ε model and standard k–ω model for two cases, namely, non-recirculating double jet at $Re = 14{,}000$ and a partially recirculating double jet at $Re = 6{,}000$ and $14{,}000$. They numerically solved the governing equations for unsteady, two dimensional flow fields. They showed that none of these models are entirely adequate in capturing the flow physics. However, the predictions by two k–ε models were superior to those by k–ω model.

Some researchers have compared the performances of the k–ε model and k–ω model for different flow configurations. The k–ε model does not accurately predict the characteristics of far wakes and mixing layers and the spreading rate of axisymmetric jets in stagnant surrounding is also overpredicted. The performance of the model can be improved by making ad hoc adjustments to the model constants. The model also has problems in swirling flows and flows with large strains (e.g., highly curved boundary layers and diverging passages). On the other hand, the k–ω model is reported to reproduce the behaviour within viscous sublayer without the need for any corrections. However, the predictions by the k–ω model are reported to be sensitive to the free-stream conditions for the free-shear flows. Table 6.1 provides a comparison of growth rates of typical free shear flows predicted by k–ε and k–ω models.

6.10 V2f Model

v2f model was proposed by Durbin (1991) and is based on the root mean square normal velocity fluctuations $\overline{v'^2}$ as the velocity scale rather than turbulence kinetic energy k. It is capable of handling the wall region without the need for the additional damping functions because the normal velocity fluctuations are known to be quite sensitive to the presence of wall and thus are like a natural damper

(Durbin 1991). The model therefore does not require a low *Re* version to treat the presence of wall. A reader may refer to Durbin and Petterson (2001) for further details of the v2f model.

The model employs four transport equations for the closure of the RANS equations, which include an equation for the turbulence kinetic energy (the equation is same as that for the standard k–ε model), the dissipation of turbulence kinetic energy (this equation is same as that used for the standard k–ε model but with a slightly modified value of the constant), a new transport equation for the normal r.m.s. velocity fluctuations and finally a transport equation for f that takes care of in-homogeneity and wall blocking effects in the transport equation for $\overline{v'^2}$. Since these two transport equations are used in conjunction with those for k and ε, this model is termed as the $k - \varepsilon - \overline{v'^2} - f$ model. There have been continuous improvements in the v2f model. The original v2f model was numerically unstable for segregated solvers.

The transport equations used can be written as (Kazerooni and Hannani 2009):

$$\frac{\partial k}{\partial t} + \overline{u_j}\frac{\partial k}{\partial x_j} = \frac{\partial}{\partial x_j}\left(\left(v + \frac{v_t}{\sigma_k}\right)\frac{\partial k}{\partial x_j}\right) + P_k - \varepsilon \tag{6.29}$$

$$\frac{\partial \varepsilon}{\partial t} + \overline{u_j}\frac{\partial \varepsilon}{\partial x_j} = \frac{\partial}{\partial x_j}\left(\left(v + \frac{v_t}{\sigma_\varepsilon}\right)\frac{\partial \varepsilon}{\partial x_j}\right) + \frac{c_{\varepsilon 1}P_k - c_{\varepsilon 2}}{T} \tag{6.30}$$

$$\frac{\partial \overline{v'^2}}{\partial t} + \overline{u_j}\frac{\partial \overline{v'^2}}{\partial x_j} = \frac{\partial}{\partial x_j}\left(\left(v + \frac{v_t}{\sigma_k}\right)\frac{\partial \overline{v'^2}}{\partial x_j}\right) - \overline{v'^2}\frac{\varepsilon}{k} + kf \tag{6.31}$$

$$f - L^2\nabla^2 f = (C_1 - 1)\frac{2/3 - \overline{v'^2}/k}{T} + C_2\frac{P_k}{k} \tag{6.32}$$

Here L and T denote turbulence length and time scales, respectively, and are given as

$$T = \min\left[\max\left[\frac{k}{\varepsilon}, C_T\sqrt{\frac{v}{\varepsilon}}\right], \frac{0.6k}{2\sqrt{3}C_\mu\overline{v'^2}S}\right] \tag{6.33}$$

$$L = C_L \max\left[\min\left[\frac{k^{3/2}}{\varepsilon}, \frac{k^{3/2}}{\sqrt{3}C_\mu\overline{v'^2}S}\right], C_\eta\frac{v^{3/4}}{\varepsilon^{1/4}}\right] \tag{6.34}$$

The term k_f in Eq. 6.31 redistributes turbulence kinetic energy from the streamwise velocity component. The eddy viscosity can be written as

$$\frac{\mu_T}{\rho} = v_T = C_\mu T\overline{v'^2} \tag{6.35}$$

The v2f model constants are given as (Durbin and Reif 2001)

$$c_\mu = 0.22, \quad c_L = 0.23, \quad c_\eta = 85, \quad c_T = 6, \quad c_1 = 0.4, \quad c_2 = 0.3$$

$$c_{\varepsilon 2} = 1.9, \quad \sigma_k = 1, \quad \sigma_\varepsilon = 1.3 \tag{6.36}$$

$$c_{\varepsilon 1} = 1.4 \left[1 + 0.045 \sqrt{\frac{k}{\overline{v'^2}}} \right]$$

The computations of conjugate heat transfer in the turbine blades is a challenging area. An engineer needs to predict the pressure loads on vanes and thermal distributions. For the later, conjugate heat transfer needs to be considered which means simultaneous and coupled solution of fluid flow and heat transfer in external hot gages and internal cooling passages and conduction within the solid. Luo and Rajinsky (2007) have performed conjugate heat transfer prediction in a turbine vane using *v2f* model, low-*Re* k–ε model and nonlinear quadratic k–ε model. They showed that v2f model provided better predictions compared to those by two other models. They suggested that the model has a good potential for turbo-machinery applications. This observation is further reinforced by the ability of the model to capture the bypass transition whose occurrence is quite common in turbo machineries.

6.11 Shear Stress Transport k–ω Model

Shear stress transport k–ω model was developed by Menter (1994) to take advantage of accurate formulation of the k–ω model in the near-wall region with the free-stream independence of the k–ε model in the far field. The SST k–ω model is similar to the standard k–ω model, but includes three refinements. (1) This model incorporates a damped cross-diffusion derivative term in the ω equation. (2) The definition of the turbulent viscosity is modified to account for the transport of the turbulent shear stress. (3) The model constants are different. There is no need for a special treatment for the viscosity affected wall region because of the low-Reynolds correction in the k–ω and k–ω SST models. The transport equations for k and ω may be written as

$$\frac{\partial k}{\partial t} + \overline{u}_j \frac{\partial k}{\partial x_j} = \frac{\partial}{\partial x_j} \left[(v + \sigma_k v_t) \frac{\partial k}{\partial x_j} \right] + P_k - \beta^* k \omega \tag{6.37}$$

$$\frac{\partial \omega}{\partial t} + \overline{u}_j \frac{\partial \omega}{\partial x_j} = \frac{\gamma P_k}{\rho v_t} - \beta \omega^2 + \frac{\partial}{\partial x_j} \left[(v + \sigma_\omega v_t) \frac{\partial \omega}{\partial x_j} \right] + 2(1 - F_1) \sigma_{\omega 2} \frac{1}{\omega} \frac{\partial k}{\partial x_j} \frac{\partial \omega}{\partial x_j} \tag{6.38}$$

The eddy viscosity is written as

$$v_T = \frac{a_1 k}{\max(a_1 \omega, SF_2)} \tag{6.39}$$

The auxiliary relations and the model constants can be written as

$$F_1 = \tanh\left(\left\{\min\left[\max\left(\frac{\sqrt{k}}{\beta^*\omega y},\frac{500v}{y^2\omega}\right),\frac{4\sigma_{\omega2}k}{CD_{k\omega}y^2}\right]\right\}^4\right) \tag{6.40}$$

$$F_2 = \tanh\left(\max\left(\frac{2\sqrt{k}}{\beta^*\omega y},\frac{500v}{y^2\omega}\right)^2\right) \tag{6.41}$$

$$P_k = \min\left(\tau_{ij}\frac{\partial\overline{u}_i}{\partial x_j},10\beta^*k\omega\right) \tag{6.42}$$

$$CD_{k\omega} = \max\left(2\rho\sigma_{\omega2}\frac{1}{\omega}\frac{\partial k}{\partial x_i}\frac{\partial\omega}{\partial x_i},10^{-20}\right) \tag{6.43}$$

The constants for example Φ are blended using the relation

$$\phi = \phi_1 F_1 + \phi_2(1 - F_1) \tag{6.44}$$

$$\gamma_1 = \frac{\beta_1}{\beta^*} - \frac{\sigma_{\omega1}k^2}{\sqrt{\beta^*}}, \quad \gamma_1 = \frac{\beta_2}{\beta^*} - \frac{\sigma_{\omega2}k^2}{\sqrt{\beta^*}}, \quad \beta_1 = 3/40, \quad \beta_2 = 0.0828, \quad \beta^* = 9/100 \tag{6.45}$$

$$\sigma_{k1} = 0.85, \quad \sigma_{k2} = 1, \quad a_1 = 0.31, \quad \sigma_{\omega1} = 0.5, \quad \sigma_{\omega2} = 0.856$$

6.12 Other Two Equation Models

We have seen some widely used two equation models and their variants in the previous sections. Some other two equation models proposed in the literature are not as popular as those already presented in the previous sections. For example, Rotta (1951) initially proposed a transport equation for turbulence length scale and subsequently proposed a transport equation for the product of turbulence kinetic energy and turbulence length scale. Saffman (1970) proposed a k–ω model.

6.13 Modifications to k–ε Model for Buoyancy Driven Flows

A plume is generated by a continuous source of buoyancy, which may be created by a continuous source of heat or concentration. In a plume the buoyancy affects both mean and turbulent quantities. A plume has a larger velocity and thermal growth rates and magnitudes of turbulent quantities are also larger compared to those in a jet (Dewan et al. 1996). The buoyancy production of turbulence also becomes important in a plume. Therefore plumes are more complex flows than the corresponding non-buoyant free-shear flows. Turbulence models of different complexities can be used to model buoyancy driven flows. However, all models need to be modified to take care of the effect of buoyancy. Here we present the mathematical modeling of a turbulent round plume using a modified version of the k–ε model. The boundary-layer assumptions can be used for the time-averaged flow in a round plume. Further the molecular diffusion term can be neglected in all the governing equations compared to the turbulent diffusion terms.

The time-averaged governing equations for the continuity, axial momentum and thermal energy may be written under the boundary layer assumptions, respectively, as

$$\frac{\partial u}{\partial x} + \frac{1}{r}\frac{\partial}{\partial r}[rv] = 0 \tag{6.46}$$

$$u\frac{\partial u}{\partial x} + v\frac{\partial u}{\partial r} = \frac{1}{r}\frac{\partial}{\partial r}\left[r\left(-\overline{u'v'}\right)\right] + g\beta(T - T_\infty) \tag{6.47}$$

$$u\frac{\partial T}{\partial x} + v\frac{\partial T}{\partial r} = \frac{1}{r}\frac{\partial}{\partial r}\left[r\left(-\overline{v't'}\right)\right] \tag{6.48}$$

Here bar over u, v and T has been omitted for simplicity and β denotes the coefficient of the volumetric expansion and is defined as increase in the volume per unit original volume per Kelvin rise in temperature. The last term on the RHS of axial momentum equation (6.47) is the buoyancy term and is responsible for initiating the fluid motion. The following modified transport equations are employed for the k–ε model

$$u\frac{\partial k}{\partial x} + v\frac{\partial k}{\partial r} = \frac{1}{r}\frac{\partial}{\partial r}\left[r\left(C_k\frac{\overline{v'^2}}{k}\right)\frac{\partial k}{\partial r}\right] + P_k + G_k - \varepsilon \tag{6.49}$$

$$u\frac{\partial \varepsilon}{\partial x} + v\frac{\partial \varepsilon}{\partial r} = \frac{1}{r}\frac{\partial}{\partial r}\left[r\left(C_\varepsilon\frac{\overline{v'^2}}{k}\varepsilon\right)\frac{\partial \varepsilon}{\partial r}\right] + c_{\varepsilon1}\frac{\varepsilon}{k}(P_k + G_k) - c_{\varepsilon2}\frac{\varepsilon^2}{k} \tag{6.50}$$

where G_k denotes the buoyancy production of turbulence kinetic energy and is given as

$$G_k = g\beta\overline{u't'} \tag{6.51}$$

G_k is in addition to the shear production of turbulence kinetic energy P_k and is approximately 30% of the total production (Dewan et al. 1996). Under the boundary-layer assumptions the shear production may be written as

$$P_k \approx -\overline{u'v'}\frac{\partial u}{\partial r} \tag{6.52}$$

Use of the gradient diffusion hypothesis $\left(\overline{u't'} = -\frac{v_t}{\mathrm{Pr}_t}\frac{\partial T}{\partial x}\right)$ is inappropriate for modeling the streamwise turbulent heat flux (and consequently the buoyancy production of turbulence) because under the boundary layer assumptions the stream-wise gradients are negligible compared to the normal gradients and therefore G_k will also turn out to be negligible whereas it is approximately 30% of the total production. Dewan et al. (1996) have proposed an improved model for turbulent axial heat flux and showed that it can work well in different situations. Their model is given as

$$\overline{u_i' t'} = -\frac{\nu_t}{\mathrm{Pr}_t}\frac{\partial T}{\partial x_i} + k_H \sqrt{k\overline{t'^2}} \tag{6.53}$$

Here k_H denotes a model constant $= 0.56$ (Dewan et al. 1996). We need to solve a transport equation for temperature fluctuations in order to obtain the buoyancy production of turbulence. The modeled transport equation for the temperature fluctuations may be written as

$$u\frac{\partial \overline{t'^2}}{\partial x} + v\frac{\partial \overline{t'^2}}{\partial y} = \frac{\partial}{\partial y}\left[\left(c_T\frac{k^2}{\varepsilon}\right)\frac{\partial \overline{t'^2}}{\partial y}\right] + P_T - \varepsilon_T \tag{6.54}$$

Here P_T and $\varepsilon_T = \alpha\overline{\frac{\partial t'}{\partial x_j}\frac{\partial t'}{\partial x_j}}$ denote the production and rate of dissipation of temperature fluctuations, respectively. According to the boundary-layer assumptions the production can be given as

$$P_T \approx -2\overline{v't'}\frac{\partial T}{\partial y} \tag{6.55}$$

We can assume that that the time scales of velocity and temperature fluctuations are proportional and the constant of proportionality is denoted as C_{T1}. This assumption results in the following simplified expression for the dissipation of temperature fluctuations.

$$\varepsilon_T = C_{T1}\varepsilon\frac{\overline{t'^2}}{k} \tag{6.56}$$

Therefore, we do not need to derive a separate transport equation for the dissipation of temperature fluctuations resulting in savings in computational cost. A general transport equation for the dissipation of temperature fluctuations may also be solved instead of the above-mentioned relation but with an increased computational cost.

The simplest way of turbulent heat flux modelling is by means of employing a constant or varying turbulent Prandtl number Pr_t. However, this is not a general approach. Karcz and Badur (2005) have presented a method of modeling turbulent diffusion of heat based on the v2f model. This concept avoids the use of the eddy viscosity and employs the transport equations for temperature fluctuations and its rate of dissipation. They showed that the model predicts well the characteristics of turbulent flow in a heated pipe.

Chow and Li (2007) considered four variants of k–ε model to compute pre-flashover stage of a compartmental fire. Building professionals are interested in such modeling for designing fire services systems. They did not consider the fire combustion and considered fire merely as a heat source. However, they modified all the governing equations (including the ones for k and ε) to account for the effect of the buoyancy. They showed that of all the models, the standard k–ε model performs the best.

6.14 Other Modifications

Turbulent flows opening up in a quiescent ambient exhibit intermittent nature in the vicinity of the ambient (e.g., the flow is not turbulent all the time but sometimes laminar and sometimes turbulent). Cho and Chung (1992) modified the k–ε model to incorporate the effects of intermittency on dissipation and solved a transport equation for the intermittency. They showed that this model is superior to the standard k–ε model for some simple free shear flows. The characteristics of the outer intermittent flow are also important for understanding the entrainment process. Dewan and Arakeri (2000) have proposed a k–ε–γ model to predict the intermittency of boundary-layers, here γ denotes the intermittency. They solved a transport equation for the intermittency. They also considered the effect of entrainment on the dissipation of turbulence kinetic energy and used a modified expression for the eddy viscosity to consider the effect of the outer intermittent flow. The modified expression for the eddy viscosity is given as

$$v_t = c_\mu \left[1 + c_{\mu g} \frac{k^3}{\varepsilon^2} \frac{1-\gamma}{\gamma^3} \frac{\partial \gamma}{\partial r} \frac{\partial \gamma}{\partial r} \right] \frac{k^2}{\varepsilon} \tag{6.57}$$

Here $c_{\mu g}$ is a model constant $= 0.10$. Clearly the effect of the intermittency is to increase the value of the eddy viscosity. In the case of the fully turbulent flow ($\gamma = 1$), this expression reduces to the expression for the eddy viscosity used in the standard k–ε model. Dewan and Arakeri (2000) showed that this model can be used in conjunction with any low-Reynolds number version of the k–ε model and is also capable of treating intermittency of different free shear flows. They showed that the model is capable of capturing the intermittency profiles of flat plate boundary-layer and thick axisymmetric boundary-layers. The value of intermittency observed over a thick axisymmetric boundary layer is one over a larger normal distance from the wall compared to that over a flat plate. This behavior is due to less constraint on the motion of eddies in a thick axisymmetric boundary-layer compared to the flat plate boundary-layer. The Reynolds shear stress also decays rapidly in the outer region of a thick axisymmetric boundary-layer compared to that in a flat plate boundary-layer as turbulence generated in the outer region sustains larger area compared to that in a flat plate (Dewan and Arakeri 2000).

In many engineering applications, such as flow along missiles, torpodeos, underwater towed cables, and during fabrication of glass and polymer fibres, the boundary layer thickness can be of the order of or greater than the body radius and thus the transverse curvature effects become important in the outer region, i.e., boundary layers are thick. The transverse curvature effects may be important in the inner region as well and to characterize these effects the radius of curvature (a) should be prescribed in the wall coordinates, i.e., $a^+ = \frac{au_\tau}{v}$. Dewan and Arakeri (1996) compared four turbulence models for wall bounded flows affected by transverse curvature: (1) zero equation model of Cebeci and Smith (1974), two low-Rek–ε models by Chien (1982) and Michelassi et al. (1993) and two-layer k–ε model of Rodi et al. (1993). They showed that though all these models are quite accurate in predicting the characteristics of the flat plate boundary layer, none of these are accurate in predicting the skin friction coefficient for thick boundary layer developing on a circular cylinder. They also showed that the velocity profile in the wall coordinates does not follow the logarithmic behaviour but shows a variable slope compared to that for the logarithmic layer. The slope depends on the radius of curvature in the wall co-ordinates a^+. Further, these models are also inaccurate in predicting the behaviour in the outer region, which shows a rather rapid decay of the Reynolds shear stress in the outer region compared to that for turbulent flow past a flat plate.

An improvement of film-cooling effectiveness has emerged as an actively studied topic in turbo-machinery applications during the last few decades and an important feature of this geometry is jet in a crossflow. This flow configuration exhibits several flow features, such as, large scale coherent structures in the flow field, streamline curvature, interaction between the jet and cross streams. Pathak et al. (2007) have shown that the standard k–ε model needs to be modified to account for the effect of the streamline curvature in order to capture the flow physics. They showed that an extra rate of strain arises due to the streamline curvature and this in turn affects the eddy viscosity.

6.15 Concluding Remarks

Two equation models with the Boussinesq assumption are widely used for predicting turbulent flows encountered in industry. Of these, the standard k–ε model is the most widely used turbulence model. The wall functions approach for treating turbulence in the vicinity of a solid wall has some limitations primarily due to the assumptions of local equilibrium between the production and dissipation of turbulence and constant shear. The wall function approach is likely to fail in the following situations: (1) Blowing/suction through the wall; (2) Large pressure gradients; (3) Strong body forces (e.g., flow near a rotating body); and (4) High three-dimensionality near the walls (e.g., strongly skewed flows). In the next chapter, we will see how we can treat turbulent flows using a more general approach compared to the Boussinesq approximation and this approach is termed as the second order moment closure modeling.

References

Cebeci T, Smith AMO (1974) Analysis of turbulent boundary layers. In: Series of Applied Mathematics and Mechanics, vol XV. Academic Press, London

Chien KY (1982) Predictions of channel and boundary layer flows with a low-Reynolds number turbulence model. AIAA J 20:33–38

Cho JR, Chung MK (1992) A k–ε–γ equation turbulence model. J Fluid Mech 237:301–322

Chow WK, Li J (2007) Numerical simulations on thermal plumes with k–ε types of turbulence models. Build Environ 42:2819–2828

Craft TJ, Launder BE, Suga K (1996) Development and application of a cubic eddy viscosity model of turbulence. Int J Heat Fluid Flow 17(2):108–115

Dewan A, Arakeri JH (1996) Comparison of four turbulence models for wall-bounded flows affected by transverse curvature. AIAA J 34:842–844

Dewan A, Arakeri JH (2000) Use of k–ε–γ model to predict intermittency in turbulent boundary-layers. ASME J Fluids Eng 122(3):542–546

Dewan A, Arakeri JH, Srinivasan J (1996) A new turbulence model for the axisymmetric plume. Appl Math Model 21:709–719

Durbin PA (1991) Near wall turbulence closure modeling without damping functions. Theor Comput Fluid Dyn 3(1):1–13

Durbin PA, Petterson RBA (2001) Statistical theory and modeling for turbulent flows. Wiley, New York

Durbin PA, Reif BAP (2001) Statistical theory and modelling for turbulent flows. Wiley, USA

Fernández JA, Elicer-Cortés JC, Valencia A, Pavageau M, Gupta S (2007) Application of linear and non-linear low-Re k–ε models in two-dimensional predictions of convective heat transfer in passages with sudden contractions. Int J Heat Fluid Flow 28:429–440

Jones WP, Launder BE (1972) The prediction of laminarization with a two equation model of turbulence. Int J Heat Mass Transf 15:301–314

Karcz M, Badur J (2005) An alternative two-equation turbulent heat diffusivity closure. Int J Heat Mass Transf 48:2013–2022

Kazerooni RB, Hannani SK (2009) Simulation of turbulent flow through porous media employing a v2f model. Sci Iran Trans B Mech Eng 16(2):159–167

Kolmogorov AN (1942) Equations of turbulent motion of an incompressible fluid. Izvestia Acad Sci USSR 6 (1&2):56–58

Launder BE, Sharma BI (1974) Application of the energy dissipation model of turbulence to the calculation of flow near a spinning disk. Lett Heat Mass Transf 1(1):131–138

Launder BE, Spalding DB (1974) The numerical computation of turbulent flow. Comput Methods Appl Mech Eng 3:269–289

Luo J, Rajinsky EH (2007) Conjugate heat transfer analysis of a cooled turbine vane using the v2f turbulence model. ASME J Turbomach 129:773–781

Menter FR (1994) Two equation eddy viscosity turbulence models for engineering applications. AIAA J 32(8):1598–1605

Michelassi V, Rodi W, Zhu J (1993) Testing a low Reynolds number k–ε turbulence model based on direct simulation data. AIAA J 31(9):1720–1723

Myong HK, Kasagi N (1990) A new approach to the improvement of k–e turbulence model for wall-bounded shear flows. JSME Int J Ser II 33(1):63–72

Nagano Y, Tagawa M (1990) An improved k–e model for boundary layer flows. Trans ASME J Fluids Eng 112:33–39

Patel VC, Rodi W, Scheuerer G (1985) Turbulence models for near wall and low Reynolds number flows: a review. AIAA J 23(9):1308–1318

Pathak M, Dewan A, Dass AK (2007) Effect of streamline curvature on flow field of a turbulent plane jet in crossflow. Mech Res Commun 34(3):241–248

Rodi W, Mansour NN, Michelassi W (1993) One equation near wall turbulence modeling with the aid of direct simulation data. Trans ASME J Fluids Eng 115(2):196–205

Rotta JC (1951) Statistiche Theorie nichthomogener Turbulenz. Z Phys 129:547–572

Saffman PG (1970) A model for inhomogoneous turbulent flow. Proc R Soc Lond A317:417–433

Sana A, Ghumman AR, Tanaka H (2007) Modification of the damping function in the k–ε model to analyse oscillatory boundary-layers. Ocean Eng 34:320–326

Shih T-H, Liou WW, Shabbir A, Zhu J (1995) A new k-ε eddy-viscosity model for high Reynolds number turbulent flows—model development and validation. Comput Fluids 24(3):227–238

Wilcox DC (1988) Reassessment of the scale determining equation for advanced turbulence models. AIAA J 19(2):248–251

Wilcox DC (2006) Turbulence modeling for CFD, 3rd edn. DCW Industries, California

Wolfstein M (1969) The velocity and temperature distribution of one-dimensional flow with turbulence augmentation and pressure gradient. Int J Heat Mass Transf 12:301–318

Yakhot V, Orszag SA (1986) Renormalization group analysis of turbulence: I basic theory. J Sci Comput 1(1):1–51

Chapter 7
Reynolds-Stress and Scalar Flux Transport Model

Abstract This chapter presents modeled forms of the Reynolds stress and scalar flux transport equations. It is shown that the modeling of several terms of the exact Reynolds stress and scalar flux transport equation is far more complex than that of the turbulence kinetic energy equation. As a result the Reynolds stress and scalar flux transport models do not always produce more accurate results compared to the two-equation models. Finally the advantages and disadvantages of such models are compared with those of the two-equation models by means of examples from the literature.

7.1 Introduction

Numerous zero-, one- and two-equation models for the closure of RANS equations are available in the literature. New models are also continuously being developed and it is not clear which one should be used for a particular flow configuration. Sometimes unsatisfactory predictions by the two-equation models can be attributed to corresponding inaccurate predictions of the Reynolds stress tensor by the eddy-viscosity models based on the Boussinesq assumption. These models are not capable of representing the highly anisotropic nature of the flow appropriately. For example, let us consider the flow field of jet in crossflow (Pathak et al. 2006). This is highly complex and anisotropic flow and therefore the use of Reynolds stress transport (RST) model, i.e., solving transport equations for different components of Reynolds stresses, can produce better results than those produced by the other two equation models. This approach is also termed as the second order moment closure (Hanjalic 2005) and we will see its details in the subsequent sections.

7.2 Modeled Equations for Reynolds Stress Transport Model

The exact transport equation for Reynolds stress tensor can be obtained by using the following mathematical operation (Wilcox 2006)

A. Dewan, *Tackling Turbulent Flows in Engineering*, 81
DOI: 10.1007/978-3-642-14767-8_7, © Springer-Verlag Berlin Heidelberg 2011

$$\overline{u_i'N(u_j)} + \overline{u_j'N(u_i)} = 0 \tag{7.1}$$

Here $N(u_i) = 0$ denotes the Navier–Stokes operator in the tensor notation and is given as

$$N(u_i) = \rho\frac{\partial u_i}{\partial t} + \rho u_k\frac{\partial u_i}{\partial x_k} + \frac{\partial p}{\partial x_i} - \mu\frac{\partial^2 u_i}{\partial x_k \partial x_k} \tag{7.2}$$

The resulting exact transport equation for the Reynolds stress tensor τ_{ij} after some adjustments can be written as

$$\frac{\partial \tau_{ij}}{\partial t} + \overline{u_k}\frac{\partial \tau_{ij}}{\partial x_k} = -\tau_{ik}\frac{\partial \overline{u_j}}{\partial x_k} - \tau_{jk}\frac{\partial \overline{u_i}}{\partial x_k} + \varepsilon_{ij} - \pi_{ij} + \frac{\partial}{\partial x_k}\left[v\frac{\partial \tau_{ij}}{\partial x_k} + C_{ijk}\right] \tag{7.3}$$

The exact pressure strain correlation term is given by $\pi_{ij} = \overline{p'\left(\dfrac{\partial u_i'}{\partial x_j} + \dfrac{\partial u_j'}{\partial x_i}\right)}$

$$\tag{7.4}$$

The exact dissipation term is given by $\varepsilon_{ij} = 2\mu\overline{\left(\dfrac{\partial u_i'}{\partial x_k}\dfrac{\partial u_j'}{\partial x_k}\right)} \tag{7.5}$

The exact turbulent transport term is given by $C_{ijk} = \overline{\rho u_i'u_j'u_k'} + \overline{pu_i'\delta_{jk}} + \overline{pu_j'\delta_{ik}}$

$$\tag{7.6}$$

The pressure-strain correlation (π_{ij}), dissipation (ε_{ij}) and turbulent transport (C_{ijk}) terms need to be modeled in order to close the exact transport equation for Reynolds stress τ_{ij}. In the subsequent sections we will briefly look into different issues involved in modeling these three terms.

7.2.1 Modeling of Turbulent Transport

This term is modeled according to the gradient diffusion assumption (Daly and Harlow 1970)

$$C_{ijk} = C_s\frac{2}{3}\frac{k^2}{\varepsilon}\left[\frac{\partial \tau_{jk}}{\partial x_i} + \frac{\partial \tau_{ik}}{\partial x_j} + \frac{\partial \tau_{ij}}{\partial x_k}\right] \tag{7.7}$$

Here $C_s \approx 0.11$ is a model constant.

7.2.2 Modeling of Pressure Strain

The modeling of pressure-strain correlation π_{ij} is a challenging task mainly because it is of about the same magnitude as the production and secondly it contains complex correlations which are difficult to measure. This term is also denoted as the pressure-strain redistribution term. Major difference between different Reynolds stress transport models is due to different models adopted for modeling of π_{ij}. The model for π_{ij} adopted by Launder et al. (1975) is one of the widely used models and is given as

$$\pi_{ij} = C_1 \frac{\varepsilon}{k}\left(\tau_{ij} + \frac{2}{3}\rho k \delta_{ij}\right) - \hat{\alpha}\left(P_{ij} - \frac{2}{3}P\delta_{ij}\right) - \hat{\beta}\left(D_{ij} - \frac{2}{3}P\delta_{ij}\right)$$

$$- \hat{\gamma}k\left(S_{ij} - \frac{1}{3}S_{kk}\delta_{ij}\right) + \left[0.125\frac{\varepsilon}{k}\left(\tau_{ij} + \frac{2}{3}k\delta_{ij}\right) - 0.015(P_{ij} - D_{ij})\right]\frac{k^{3/2}}{\varepsilon n}$$

The model constants are given as (Launder 1989)

$$\hat{\alpha} = (8 + C_2)/11, \hat{\beta} = (8C_2 - 2)/11, \hat{\gamma} = (60C_2 - 4)/55, C_1 = 1.8, C_2 = 0.6 \tag{7.9}$$

The auxiliary relations are given as

$$P_{ij} = \tau_{im}\frac{\partial \overline{u}_j}{\partial x_m} + \tau_{jm}\frac{\partial \overline{u}_i}{\partial x_m}, \quad D_{ij} = \tau_{im}\frac{\partial \overline{u}_m}{\partial x_j} + \tau_{jm}\frac{\partial \overline{u}_m}{\partial x_i}, \quad P = \frac{1}{2}P_{kk} \tag{7.10}$$

7.2.3 Modeling of Dissipation

The dissipation of Reynolds stress is also a tensor and this is usually modeled as

$$\varepsilon_{ij} = \frac{2}{3}\varepsilon\delta_{ij} \tag{7.11}$$

where ε denotes the scalar dissipation rate of turbulence kinetic energy $\varepsilon = \nu\overline{\frac{\partial u_i'}{\partial x_k}\frac{\partial u_i'}{\partial x_k}}$ and δ_{ij} the Kronecker delta. $\delta_{ij} = 1$ if $i = j$ and $\delta_{ij} = 0$ if $i \neq j$. The use of this assumption essentially avoids the need for employing a dissipation transport equation for each component of Reynolds stress tensor τ_{ij} and results in a reduction in the number of transport equations to be solved and thus the computational cost. In most second moment closures, the equation for the dissipation (ε) is similar to that employed for the standard k-ε model. As we know the dissipation can be anisotropic close to the wall and therefore some researchers have treated dissipation anisotropically without solving a transport equation for each component (see for example, Hanjalic and Launder 1976).

7.3 Exact Transport Equation for Scalar Flux

In the same way a transport equation for turbulent scalar flux can be derived (Rossi et al. 2009) and is written as

$$\frac{D\overline{\theta' u_i'}}{Dt} = -\overline{u_i u_j}\frac{\partial \theta}{\partial x_j} - \overline{\theta u_j}\frac{\partial U_i}{\partial x_j} - \varepsilon_{i\theta} - \phi_{i\theta} + d_{ik\theta} \tag{7.12}$$

The first two terms on the RHS of Eq. 7.12 denote the scalar flux production and the remaining three terms can be given by the following expressions

$$\text{Dissipation } \varepsilon_{i\theta} = (D + v)\overline{\left(\frac{\partial u_i}{\partial x_k}\frac{\partial \theta}{\partial x_k}\right)} \tag{7.13}$$

$$\text{Pressure scalar gradient correlation } \phi_{i\theta} = \overline{\frac{p'}{\rho}\frac{\partial \theta}{\partial x_i}} \text{ and}$$

$$\text{Turbulent transport } d_{ik\theta} = -\frac{\partial}{\partial x_l}\left(\overline{u_i' u_i' c'} + \overline{\frac{p' c'}{\rho}}\delta_{ij}\right) \tag{7.14}$$

7.4 Boundary Conditions

Here we present typical boundary conditions that can be adopted for solving turbulent flow and temperature fields using the Reynolds stress and scalar flux transport model.

Inlet: At the inlet, distributions of all components of Reynolds stress tensor and dissipation of turbulence kinetic energy ε along with mean velocity components (two components in two-dimensional time mean flow and three components in three-dimensional time mean flow) and mean temperature should be specified. If it is a two-dimensional mean flow, three components of the Reynolds stress tensor $(-\rho\overline{u'^2}, -\rho\overline{v'^2}, -\rho\overline{u'v'})$ need to be specified and for a three-dimensional mean flow, six components $(-\rho\overline{u'^2}, -\rho\overline{v'^2}, -\rho\overline{w'^2}, -\rho\overline{u'v'}, -\rho\overline{u'w'}, -\rho\overline{v'w'})$ need to be specified. Only one scalar dissipation (ε) needs to be specified if dissipation is modeled using Eq. 7.11.

If the detailed inlet boundary condition information is not available the following expressions may be used

$$k = \frac{3}{2}(U_{\text{ref}} T_i)^2, \quad \varepsilon = C_\mu^{3/4} k^{3/2}/l, \quad l = 0.07L, \overline{u_1'^2} = k,$$

$$\overline{u_2'^2} = \overline{u_3'^2} = \frac{1}{2}k, \quad \overline{u_i' u_j'} = k \; (i \neq j) \tag{7.15}$$

Here U_{ref} denotes some reference velocity at the inlet, T_i the turbulence intensity and L the length scale of the flow configuration.

Outlet: The stream-wise gradients of the mean and turbulent flow variables are taken to be zero at the outlet and the pressure is taken equal to the atmospheric pressure.

Symmetry plane: At the symmetry plane $\partial u_i/\partial n = 0$, $\partial T/\partial n = 0$, $\partial \tau_{ij}/\partial n = 0$ and $\partial \varepsilon/\partial n = 0$ where n denotes the normal to the symmetry plane.

Free-stream: At a free stream, turbulence is usually zero, i.e., $\tau_{ij} = 0$ and $\varepsilon = 0$. The mean velocity, temperature and pressure are equal to the corresponding ambient values.

7.5 Treatment of Solid Walls

A standard Reynolds stress and scalar flux transport (RSSFT) model is valid for the fully turbulent flows and therefore not valid in the presence of solid walls where turbulence Reynolds number is low. In the near wall region, the molecular viscosity affects the generation, destruction and transport of the Reynolds stresses. As for the standard k-ε model, we can either use the wall functions or low Reynolds version of the RSSFT transport model. In the second approach, suitable modifications need to be introduced to the model to make it applicable in the vicinity of a solid wall.

The wall-function type boundary conditions for RSSFT model are similar to those used for the k-ε model. At high value of Reynolds number, the mean velocity and temperature at the first grid point close to the wall are obtained from Eqs. 6.9 and 6.11, respectively. The near wall Reynolds stress values are computed from the expressions such as

$$\tau_{ij} = \overline{u_i' u_j'} = c_{ij} k \tag{7.16}$$

where c_{ij} denote some constants and their values are obtained from the measurements. The values of k and ε at the first grid point are obtained from Eq. 6.10.

The wall functions attempt to remove the shortcomings of the RSSFT model in the near wall region. In low Re versions of RSSFT model, wall damping functions to adjust the constants of the ε-equation and a modified dissipation rate provide realistic modeling near the solid walls. The model of Hanjalic et al. (1999) is one of the widely used low-Re version of RST model.

Gerolymos et al. (2002) presented an interesting method of obtaining the near wall computations using a near wall RST model with the computing time nearly 30% more than that using the standard k-ε model. Their model is independent of the normal distance from the wall. They considered three complex turbomachinery computations involving flow separation and included a new term to account for the Corioli's redistribution effect of Reynolds stresses.

They suggested that the model needs to be further improved to predict the reattachment.

7.6 Features of Reynolds-Stress and Scalar Flux Transport Model

This is the most complex turbulence model for the closure of Reynolds-averaged Navier–Stokes and energy equations. The RSSFT model closes the Reynolds-averaged Navier–Stokes equations by solving transport equations for the Reynolds stresses and scalar fluxes, together with an equation for the dissipation rate. This means that four additional transport equations (for two Reynolds normal stresses, one Reynolds shear stress and one for dissipation) are required in a two-dimensional mean flow and seven additional transport equations (three Reynolds shear stresses, three Reynolds normal stresses, and one for dissipation) must be solved in a three-dimensional mean flow. This model accounts for the effects of streamline curvature, swirl, rotation, and rapid changes in strain rate in a more physical manner than that by one-equation and two-equation models. It has a good potential to provide accurate predictions of complex flows. It is used for computing cyclone flows, swirling flows in combustors, rotating flow passages, and the stress-induced secondary flows in ducts. Mixing length and k-ε models assume that eddy viscosity is isotropic, i.e., the ratio of Reynolds-stress and mean rates of strain is same in all directions. This assumption may not be valid in many situations and therefore the use of RSSFT model is advisable in such situations. It is to be noted the use of a RSSFT model significantly increases the computational cost due to the employment of seven additional partial differential equations. However, RSSFT models are not as widely used as the k-ε model. Tummers et al. (2007) performed an interesting comparison of the standard k-ε model and Reynolds stress transport model of Hanjalic et al. (1999) for the predictions of near wake of a flat plate with and without the adverse pressure gradient. They showed that none of the two models correctly predict the mean flow reversal near the centerline.

7.7 Algebraic Stress and Scalar Flux Models

If the convective and diffusive transport terms in the transport equations for Reynolds stress tensor are removed or modeled, the Reynolds stress equations reduce to a set of algebraic equations (Rodi 1976). This can be achieved by assuming that the sum of the convection and diffusion terms of the Reynolds stresses is proportional to the sum of the convection and diffusion terms of turbulent kinetic energy and this results in

$$\frac{D\overline{u_i'u_j'}}{Dt} - D_{ij} \approx \frac{\overline{u_i'u_j'}}{k}\left(\frac{Dk}{Dt} - [k \text{ diffusion terms}]\right) \tag{7.17}$$

The Reynolds stress tensor using the algebraic stress model is given as

$$\tau_{ij} = \overline{u_i'u_j'} = \frac{2}{3}k\delta_{ij} + \left(\frac{C_D}{C_1 - 1 + P/\varepsilon}\right)\left(P_{ij} - \frac{2}{3}P\delta_{ij}\right)\frac{k}{\varepsilon} \tag{7.18}$$

Here C_D and C_1 denote the model constants.

It is clear that the expression for the Reynolds stress is similar to that for the one using the eddy viscosity. A set of six simultaneous algebraic equations for the six unknown Reynolds stresses τ_{ij} can be solved if k and ε are known. Therefore, an algebraic stress model can be applied in conjunction with the standard k-ε model equations. Computationally it is the cheapest method to account for Reynolds stress anisotropy and it combines the advantages of RST model with the economy of k-ε model. It is successfully applied for isothermal and buoyant thin shear layer flows. Recently Rossi et al. (2009) have compared the performance of algebraic heat flux models by predicting the scalar dispersion from a point source over a wavy wall and showed that these models provide accurate predictions in complex situations.

7.8 Examples

Flow normal to a circular cylinder is relevant to several practical applications, e.g., offshore risers, bridge piers, chimneys, towers, cables, antennas, etc. Typically this flow is highly three-dimensional and is characterized by unsteadiness. Presence of another cylinder in the wake of an upstream cylinder adds to the complexity. Such situations are encountered in heat exchangers, cooling systems for nuclear power plants, buildings and power transmissions (Said et al. 2008). The flow behaviour in such a situation is a strong function of the separation between the cylinders. Said et al. (2008) simulated three-dimensional unsteady flow field using a Reynolds stress model of Lien and Leschziner (1994) and compared their predictions with their measurements using PIV for Reynolds number varying from 8.5×10^3 to 6.4×10^4. They used a fine mesh with the first point located at $y^+ = 1$. They used the wall function approach and the velocity at the first grid point was obtained from the linear expression in the viscous sublayer, i.e., $u^+ = y^+$. They investigated the near wake, mid wake and far wake of the cylinder. A good agreement was obtained between the predictions of mean and turbulent quantities and the experimental data. The computed flow field showed a vortex downstream of the first cylinder and vortices in the intermediate region. A separation flow was observed on the free end of the downwind cylinder model. They showed that the effect of the upwind cylinder on the downwind diminishes with increasing spacing between the two.

Systems involving complex turbulent flows with separation, subsequent reattachment and stagnation regions require accurate modeling as they have a significant influence on the fluid flow and heat transfer characteristics. The main difficulties in Reynolds stress transport models are modeling of the pressure-strain terms and the near wall turbulence (Jia et al. 2007). The near wall modeling needs low-Re models which are important for heat transfer predictions. Many available low-Re Reynolds-stress transport models show a problem for flows with separation and reattachment, such as backward facing step (Jia et al. 2007). Jia et al. (2007) proposed a new RST model aimed for engineering applications with the consideration of near wall turbulence. The model employs the SSG pressure-strain term, the ω-equation and the SST model for the shear stresses in the near wall region. They selected these models on the merits that Speziale et al. (1991) (SSG) RST model performs well in the fully turbulent region and the shear stress transport (SST) model is a proper two-equation model that performs well for flows with adverse pressure gradient. They selected a function for the blending of the RST model and SST model. By this blending the unphysical bulge at the reattachment point is eliminated. They concluded that since the anisotropy is naturally predicted for the impinging case, the model accurately predicts both mean and turbulent flow fields.

Turbine blade passages are subjected to Coriolis and centrifugal buoyancy forces. Accurate predictions of the flow and local heat transfer in the cooling passages are important to maintain blade temperatures and thus enhance the life of the components. Chen et al. (2000) investigated flow in an unsteady, rotating flow in a two pass square channel with 180° bend at Reynolds number 25000 using a near wall version of the second moment closure model. They also employed two-layer isotropic eddy-viscosity model of Chen and Patel (1988) and compared the predictions with the experimental data of Wagner et al. (1991). The two-layer model uses the standard k-ε model in the outer region and one equation k-l model in the inner region. They showed that the due to the above-mentioned effects the flow is highly anisotropic and the Reynolds stress transport model accurately captures the flow features.

7.9 Concluding Remarks

The second order closure is the most complex and physically the most realistic among all other closure options that can be used. However, the additional complexity does not necessarily mean that the predictions using a Reynolds stress and scalar flux transport model will be more accurate compared to those obtained using other simpler models. This anomaly arises due to the uncertainty involved in modeling complex double and triple correlations in the transport equation for the Reynolds stress tensor and turbulent scalar flux. In other words, the modeled form of the Reynolds stress and turbulent scalar flux transport equations may not accurately represent the physics of the flow. In the next chapter, we will present another significantly different approach, i.e., solution of instantaneous, three-dimensional flow field using large eddy simulation and direct numerical simulation.

References

Chen HC, Patel VC (1988) Near-wall turbulence models for complex flows including separation. AIAA J 26(6):641–648

Chen HC, Jang YJ, Han JC (2000) Computation of heat transfer in rotating two-pass square channels by a second-moment closure model. Int J Heat Mass Transf 43:1603–1616

Daly BJ, Harlow FH (1970) Transport equations in turbulence. Phys Fluids 13:2634–2649

Gerolymos GA, Neubauer J, Sharma VC, Vallet I (2002) Improved prediction of turbomachinery flows using near-wall Reynolds-stress model. Trans ASME J Turbomach 124:86–99

Hanjalic K (2005) Will RANS survive LES? A view of perspectives. ASME J Fluids Eng 127(5):831–839

Hanjalic K, Launder BE (1976) Contribution towards a Reynolds stress closure for low-Reynolds-number turbulence. J Fluid Mech 74(4):593–610

Hanjalic K, Hdzic I, Jakirlic S (1999) Modelling of turbulent wall flows subjected to strong pressure variations. ASME J Fluids Eng 121:57–64

Jia R, Sundén B, Faghri M (2007) A new low Reynolds-stress transport model for heat transfer and fluid in engineering applications. Trans ASME J Heat Transf 129:434–440

Launder BE (1989) Second-moment closure: present and future? Int J Heat Fluid Flow 10(4):282–300

Launder BE, Reece GJ, Rodi W (1975) Progress in the development of Reynolds stress turbulence closure. J Fluid Mech 68(3):537–566

Lien FS, Leschziner MA (1994) Assessment of turbulent transport models including non-linear RNG eddy viscosity formulation and second-moment closure. Comp Fluids 23:983–1004

Pathak M, Dewan A, Dass AK (2006) Computational prediction of a slightly heated turbulent rectangular jet discharged into a narrow channel crossflow using two different turbulence models. Int J Heat Mass Transf 49(21–22):3914–3928

Rodi W (1976) A new algebraic relation for calculating Reynolds stresses. ZAMM 56:219

Rossi R, Philips DA, Iaccarino G (2009) Numerical simulation of scalar dispersion in separated flows using algebraic flux models. Turbul Heat Mass Transf 6:1–12

Said NM, Mhiri H, Bournot H (2008) Experimental and numerical modelling of the three-dimensional incompressible flow behaviour in the near wake of circular cylinders. J Wind Eng Ind Aero 96:471–502

Speziale CG, Sarkar S, Gatski S (1991) Modeling the pressure-strain correlation of turbulence: an invariant dynamical systems approach. J Fluid Mech 227:245–272

Tummers MJ, Hanjalic K, Passchier DM, Henkes RAWM (2007) Computations of a turbulent wake in a strong adverse pressure gradient. Int J Heat Fluid Flow 28:418–428

Wagner JH, Johnson BV, Kopper FC (1991) Heat transfer in rotating serpentine passage with smooth walls. J Turbomach 113(3):321–330

Wilcox DC (2006) Turbulence modeling for CFD, DCW Industries Inc, La Canada, CA 91011, USA

Chapter 8
Direct Numerical Simulation and Large Eddy Simulation

Abstract This chapter presents important issues that one needs to consider in order to perform direct numerical simulation (DNS) and large eddy simulation (LES) of turbulent flows. We also present important similarities and differences between DNS and LES. We show that DNS though potentially a powerful tool cannot be used as a design tool for engineering applications and show LES has a good potential as a design tool. We present different subgrid scale models that can be used to perform LES with the advantages and disadvantages of each one. We present challenges in performing LES and DNS especially in the presence of wall and in applications dealing with convective heat transfer. Different ways of treating a solid wall by using suitable models in LES are also presented in this chapter.

8.1 Introduction

In direct numerical simulation (DNS), the governing partial differential equations for instantaneous, three-dimensional turbulent flow are numerically solved and therefore no modeling is required. In DNS all scales ranging from the smallest scales, where the dissipation of the turbulence kinetic energy into thermal energy takes place (termed as the Kolmogorov length scale), up to the largest scales (typically defined by the characteristic length of the flow configuration being considered) are resolved both in space and time. Therefore DNS requires enormous computational efforts. DNS studies are still limited to the flow problems with small Reynolds numbers and simple geometries. In large eddy simulation (LES), we also solve for instantaneous, three-dimensional flow field, but as the name suggests, only the large scales are resolved and the effects of the smallest scales are modeled. This is because the large scale eddies are known to be largely affected by the boundary conditions and therefore should be computed, but small scales are largely independent of the flow geometry, tend to be homogeneous and

A. Dewan, *Tackling Turbulent Flows in Engineering*,
DOI: 10.1007/978-3-642-14767-8_8, © Springer-Verlag Berlin Heidelberg 2011

isotropic, irrespective of geometry being considered and therefore can be easily modeled. Although LES requires less computational efforts than that for DNS, the computational requirements of the LES remain rather high for the optimization of complicated processes where a large number of computations need to be performed.

Since one deals with three-dimensional, instantaneous flow fields in DNS and LES, it results in two major difficulties: (a) the generation of accurate initial conditions at all points within the computational domain and (b) boundary conditions at all instants for all flow variables for three-dimensional flow field corresponding to a particular flow configuration. These difficulties are due to the fact that in engineering applications one is usually interested in a time-averaged flow field and therefore time averaged flow field can be easily specified in terms of initial and boundary conditions. However, it is difficult to prescribe initial and boundary conditions for three-dimensional, instantaneous flow field. The problem is further compounded by the fact that the initial conditions affect the flow for several eddy turnover times (given by $l/\sqrt{k}$ where l denotes a typical length scale and k turbulence kinetic energy). Eddy turn over time is defined as the typical time for an eddy to turn around and interact with its surroundings. Therefore the prescription of inaccurate initial conditions can result in unphysical computed flow field.

DNS and LES also produces huge computational data which contains all computed flow properties at all grid points (with extremely fine grid resolutions) within the computational domain and at all instants (with extremely small time steps) both of which can be several millions in number. The data needs to be stored to obtain the time-averaged flow field and therefore the management of huge computed data is an important issue in DNS and LES. No such problems exist while employing the RANS equations to compute the mean flow field. In the subsequent sections we will consider in detail issues related with DNS and LES.

8.2 Direct Numerical Simulation

DNS is the most accurate approach to turbulent flow simulation. In DNS we numerically solve the instantaneous, three-dimensional, Navier–Stokes equations without any averaging or approximation. It requires no modeling, but it demands resolution in both space and time from the largest scales to the smallest scales at the beginning of the dissipation scales. The numerical discretization errors can be estimated and controlled. To limit discretization errors in the solution, it is essential to use higher order numerical schemes. The eddies of all length scales varying from the largest, which are about the size of physical dimensions, down to the smallest Kolmogorov scale need to be resolved. The length (η), time (τ) and velocity scales (v) of the smallest eddies can be characterized based on the dimensional reasonings, respectively, as

$$\eta = \left(\frac{v^3}{\varepsilon}\right)^{1/4}, \ \tau = \left(\frac{v}{\varepsilon}\right)^{1/2}, \ v = (v\varepsilon)^{1/4} \tag{8.1}$$

where v and ε denote the molecular kinematic viscosity and dissipation, respectively.

In DNS, the cost of simulation scales as Re_t^3, where Re_t denotes the turbulence Reynolds number. Simulations can need up to several hundreds of hours of computations depending on the geometry being considered and the value of Re. Therefore DNS is usually limited to simple geometries and small Reynolds numbers. DNS data contains detailed information about the flow and sometimes it is too much informative for any engineering purpose. Since DNS is expensive it cannot be used a design tool. DNS research includes development of detailed understanding of flow physics, mechanisms of turbulence production, energy transfer and dissipation in turbulent flows. This information can be used for the development and validation of RANS based turbulence models presented in the previous chapters.

8.3 Large Eddy Simulation

Turbulent flows are characterized by eddies with a wide range of length and time scales. The largest eddies are typically comparable in size to the characteristic length of the mean flow. The smallest scales are responsible for the dissipation of turbulence kinetic energy. In LES, large eddies are directly computed, while small eddies are modeled (Fig. 8.1). Momentum, mass, energy, and other passive scalars are transported mostly by large eddies. Large eddies are problem-dependent, i.e., they are dictated by the geometries and boundary conditions of the flow being

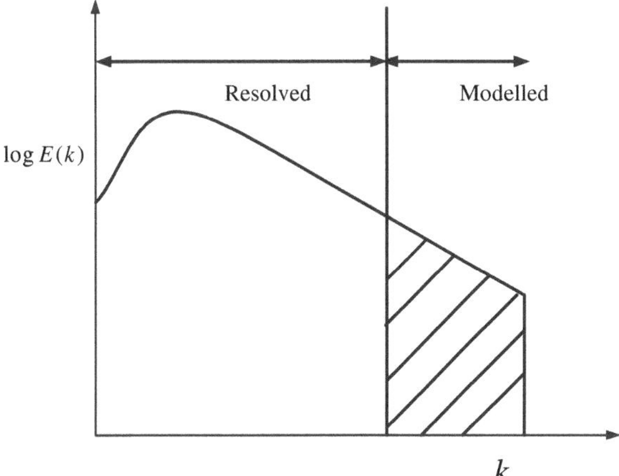

Fig. 8.1 A typical energy spectrum captured by a large eddy simulation

considered. Small eddies are less dependent on the geometry, tend to be isotropic, and are consequently universal compared to large eddies. Therefore the possibility of finding a universal turbulence model is much higher for small eddies compared to large eddies, irrespective of the geometry being considered. Resolving only the large eddies allows one to use coarser mesh sizes and larger times-steps in LES than that in DNS. LES still requires substantially finer meshes than those typically used for RANS calculations. Large eddy simulation (LES) thus falls between DNS and RANS in terms of computational cost. LES has to be run for a sufficiently long flow-time to obtain stable statistics of the flow being modeled. The computational cost involved with LES is normally orders of magnitudes higher than that for steady RANS calculations in terms of memory (RAM) and CPU time. Therefore, high-performance computing (e.g., parallel computing) is a necessity for LES, especially for industrial applications. LES is believed to be more accurate and reliable than RANS for flows in which large-scale unsteadiness is dominant such as the flow over bluff bodies.

As we have already seen, LES requires modeling of part of the inertial sub range and into the beginning of the dissipation scales (Fig. 8.1). LES resolves the large flow structures by selecting appropriate filter function. The short wave information, lost through filtering, is called subgrid component. The modeling of small scale unresolved stresses is done using a subgrid scale model (SGS model). The filtered Navier–Stokes equations can be written in the tensor notation for instantaneous, three-dimensional flow as

$$\frac{\partial \overline{u_i}}{\partial t} + \frac{\partial (\overline{u_i u_j})}{\partial x_j} = -\frac{\partial \overline{p}}{\partial x_i} + v\frac{\partial}{\partial x_j}\left(\frac{\partial \overline{u_i}}{\partial x_j} + \frac{\partial \overline{u_j}}{\partial x_i}\right) + \frac{\partial \tau_{ij}}{\partial x_j} \tag{8.2}$$

The filtered thermal energy equation is

$$\frac{\partial \overline{\theta}}{\partial t} + \overline{u_j}\frac{\partial \overline{\theta}}{\partial x_j} = -\frac{\partial q_i}{\partial x_i} \tag{8.3}$$

The source term has not been considered in Eq. (8.3). Here τ_{ij} denotes the SGS stress tensor and is defined as

$$\tau_{ij} = \overline{u_i u_j} - \overline{u_i}\,\overline{u_j} \tag{8.4}$$

Likewise the SGS heat flux tensor in the thermal energy equation is defined as

$$q_i = \overline{u_i \theta} - \overline{u_i}\,\overline{\theta} \tag{8.5}$$

The SGS stress term contains unknown velocity correlation which has to be modeled. This tensor describes the interaction between the large resolved grid scales and the small unresolved SGS. In Sect. 8.4 we will study some models that can be used to take into account the subgrid scales, with their advantages and disadvantages.

LES can be characterized into different categories depending on the fraction of the energy spectrum is resolved. This in turn depends on flow configuration. For example, in a wall bounded flow, the smallest scales are much smaller in the vicinity of a solid wall than those far away. Therefore, in such a situation it is possible to solve for a significant fraction of energy away from the wall with a reasonably fine mesh, whereas in the vicinity of a solid wall some suitable turbulence model needs to be used as its accurate resolution requires an extremely fine mesh close to the wall. Several models have been proposed in the literature for the treatment of the wall. For fine mesh the wall shear stress obtained from the laminar stress–strain relationship is used. For coarse mesh the logarithmic law-of-the-wall is used with a blending at the buffer region. The Werner and Wengle (1991) wall functions are also widely used. Two approaches used to treat solid walls in LES can be broadly classified as

(1) Use the wall models based on equilibrium laws (i.e., logarithmic law) presented in Sect. 8.7.1.
(2) Two layer models proposed by Balaras et al. (1996) mostly employ RANS equations to provide an estimate of the wall shear stress. They use a one-dimensional grid between the wall and first LES computation node and solve RANS equations on the inner mesh. This is also termed as thin boundary-layer equation (TBLE) model. There is an exchange of information between outer LES mesh and inner RANS mesh so as to avoid discontinuities at the interface. The pressure gradient is assumed to be constant in the wall normal direction and therefore saving in the computational effort by avoiding a numerical solution of the Poisson equation.

We will see details of the two approaches in Sect. 8.7.

8.4 Subgrid Scale Models for LES

Several models for the subgrid scales exist in the literature. We will discuss here details of different widely used models with their advantages and disadvantages.

8.4.1 Smagorinsky SGS Model

This is the simplest and earliest SGS model was proposed by Smagorinsky (1963). It is based on the gradient diffusion hypothesis of Boussinesq and the subgrid stress is given as

$$\tau_{ij} = 2v_{\mathrm{sgs}}\overline{S_{ij}} \tag{8.6}$$

Similarly subgrid stress in the thermal energy equation can be modeled as

$$\tau_{\theta i} = 2\frac{v_{\mathrm{sgs}}}{\mathrm{Pr}_t}\frac{\partial T}{\partial x_i} \tag{8.7}$$

where v_{sgs} denotes the subgrid scale eddy viscosity and $\overline{S_{ij}}$ the strain rate of the resolved flow field. The latter is given as

$$\overline{S}_{ij} = \frac{1}{2}\left(\frac{\partial \overline{u}_i}{\partial x_j} + \frac{\partial \overline{u}_j}{\partial x_i}\right) \tag{8.8}$$

The subgrid scale eddy viscosity v_{sgs} is a scalar and is defined using the Smagorinsky model as

$$v_{sgs} = (K_s\Delta)^2\left(2S_{ij}S_{ij}\right)^{1/2} \tag{8.9}$$

Here Δ denotes the width of the filter which is generally taken equal to the mesh size, K_s the Smagorinsky constant and v_{SGS} the eddy viscosity. For statistically stationary isotropic turbulence, with Δ in the inertial range, the value of K_s is found to be equal to 0.16. However for other flows the value of K_s is not constant and it depends on the type of flow being studied. For example, in the vicinity of a solid wall the eddy viscosity should reduce to the zero right at the wall ($y = 0$). Therefore a damping function needs to be added to the eddy viscosity. Most researchers use a van Driest damping function like the one used for RANS models.

$$K_s = K_{so}(1 - e^{-n^+/A^+})^2 \tag{8.10}$$

Here n^+ is the non-dimensional distance from the wall, A^+ a constant is equal to 25 and K_{so} the value of eddy viscosity away from the wall.

It is known that in a time-averaged turbulent flow field, energy transfer from large eddies to smaller eddies only takes place. However, in an instantaneous flow reverse cascade takes place, i.e., transfer of energy from small scales to large scales, and this is termed as the back scatter of energy. The Smagorinsky model, however, cannot capture this behaviour. Further it is known that SGS stress and the resolved strain rate tensors are not always aligned, as assumed in the Smagorinsky model. Thus the Smagorinsky's model is not without its limitations.

8.4.2 Dynamic SGS Model

Dynamic SGS model may be considered as a modification of Smagorinsky's model (Germano et al. 1991). In a dynamic subgrid scale model, the Smagorinsky's constant is computed locally at each time step during the LES calculation, based on the resolved flow fields, thus requiring no user inputs as in the case of the Smagorinsky model. This results in accurate representation of the flow field especially since the field changes considerably within the computational domain in both space and time. However, this is achieved with increased computational effort.

8.4.3 Scale Similarity SGS Model

Scale similarity SGS model was proposed to overcome the drawbacks of the eddy viscosity type models. The scale similarity model of Bardina et al. (1980) is based on the idea that the interaction between the resolved (computed) and unresolved (filtered) quantities is via the smallest resolved and the largest unresolved quantities and therefore depends on the length scale associated with the filter. These models do not assume that the resolved strain rate tensor is aligned with the SGS stress tensor and that there is a local balance between the production and dissipation of SGS kinetic energy.

8.5 DNS vis-à-vis LES

The first major similarity between the DNS and LES is that in both instantaneous and three-dimensional flow fields are obtained. Thus both are closest to the raw form of turbulent flow, i.e., without any simplification and time averaging. The major differences between the two can be summarized as

(1) DNS employs larger number of grid points and time steps compared to LES.
(2) Theoretically DNS is the most accurate approach. The accuracy of DNS computations is directly related to the numerical errors. However, in addition to numerical errors, a LES also depends on how accurately a SGS model captures the flow features. Therefore LES is somewhat approximate in nature compared to DNS.
(3) The near wall structures in a turbulent flow are extremely important. However, an accurate resolution of these structures both spatially and temporally is quite expensive and therefore computational requirements for both LES and DNS in the vicinity of a solid wall are quite large. However, the computational requirements for LES are somewhat less than those for DNS. We will consider the treatment of wall in LES in Sect. 8.7.
(4) The grid in a LES can be larger than the Kolmogorov scale, whereas the one in DNS has to be of the same size as the Kolmogorov scale.

8.6 Detached Eddy Simulation and Hybrid Models

Detached eddy simulation (DES) was proposed by Spalart et al. (1997) to combine the best aspects of RANS computations and large eddy simulation in a single solution setup. DES was primarily designed for separated flows. As we have already seen an accurate LES of such flows will require extremely large computational effort and we have already seen that the RANS computations may not be reliable after separation. As already mentioned, DNS of such flows, though potentially quite accurate, is quite expensive. The basic idea in a detached eddy simulation is to have the flow solver sensitive to the grid spacing. RANS computations are initiated in the regions near solid boundaries (wall) and where the turbulent length scale is less than the maximum grid dimension.

DES was originally proposed by Spalart by replacing the function d of the Spalart and Allmaras (1992) one equation model by a modified distance function

$$d' = \min(d, C_{\mathrm{DES}}\Delta) \tag{8.11}$$

Here C_{DES} denotes a constant, Δ the largest size of the grid being considered and d the distance from the wall. In summary as the turbulent length scale exceeds the grid dimension, the regions are solved using the LES mode and when it is smaller the computations are performed using the RANS equations. In DES, the Spalart and Allamaras (1992) model is used for both RANS and LES computations. DES has primarily been used in low-speed flows and is suitable for separated flows. However, DES is not without limitations. Switching between RANS and LES based on the mesh size introduces some uncertainty. Further the use of Spalart and Allmaras (1992) model, which is primarily meant for external aerodynamic flows, to other flows is also not without doubt.

Another interesting option is termed as the hybrid approach and involves the use of both LES and RANS equations. In this approach a rather coarse LES is applied in outer region and RANS with a fine mesh is performed in the wall region. An interface between the two needs to be chosen and a suitable matching condition needs to be applied there. Therefore an important issue is what should be the right interface condition because the nature of the governing equations as well as the computed quantities used on the either sides of the interface in RANS and LES are different. An equality of the total stress or total viscosity is a common choice of the condition at the interface (Hanjalic 2005). Usually the eddy viscosity obtained from the RANS computations is much larger than that obtained from LES and therefore some damping needs to be introduced in the former and this introduces some arbitrariness to the flow solver. This feature also introduces some kink in the velocity profile at the interface. Some proposals have been suggested in the literature to remove this problem. This problem is further compounded if the interface is located in the close vicinity of solid wall (y^+ approximately equal to 100). The magnitude of the problem can be reduced if the interface is placed farther away from the wall.

8.7 Treatment of Walls in LES

Performing either DNS or LES for a flow in the presence of a solid wall is a challenging task. This is because the flow structure is very fine in the vicinity of a solid wall with an extremely complex topology. For example, in the streamwise (x) direction the typical size of structures may be approximately 50 wall units and in spanwise (z) and normal (y) directions these could be approximately 20 and 1 wall units, respectively. A wall unit is defined as the separation non-dimensionalized using the friction velocity (u_τ) and molecular viscosity (υ). Reynolds (1990) showed that computational costs associated with LES in the presence of a solid wall grow as $Re^{2.4}$ if the no slip condition at the solid wall is applied and all structures are accurately resolved. In comparison, a LES requires grid density to increase as $Re^{0.4}$ in regions away from a solid wall, which is not a coarse resolution. If, however, a coarse grid is used, meaning thereby, that adequate numbers of grids are not located in the wall region, then the computed results may not be accurate. Clearly such situations require the use of a highly anisotropic grid in the vicinity of a solid wall.

Piomelli and Balaras (2002) estimated that for a Reynolds number of 1×10^6 nearly 90% of the nodes should be used to compute the inner region which constitutes nearly 10% of the boundary-layer region. Many proposals exist in the literature to reduce the mesh requirements in the near wall region while performing LES so that a coarser mesh can be used there. These approaches can be divided in two categories. One way to avoid the need for a fine mesh near a wall is to use the wall functions approach like the one used for the RANS equations. The only difference is that the time-averaged wall functions are applied to the instantaneous flow field. However, this approach shares the weaknesses of the wall functions for the RANS equations, i.e., not being universally applicable to a wide range of flows. The other approach is the hybrid approach, where the RANS equations are solved in the near wall region and in the outer region LES is performed. We have already seen the hybrid approach in Sect. 8.6. We will see details of the first approach in the next section.

8.7.1 Wall Models That Use Equilibrium Laws

The wall model of Werner and Wengle (1991) is one of earliest ones. It uses the expression $\vec{\tau}_w = f(|\vec{v}_{\tan}|)\,\overrightarrow{v}_{\tan}$ where the function f is based on the law of the wall applied to instantaneous quantities rather than to the standard time averaged quantities considered while treating the RANS equations. A major limitation of this approach is that a straight forward conversion of the law of the wall from the time averaged quantities to the instantaneous quantities is questionable. Further the standard wall laws for time-averaged flow properties are applicable to the attached flows without significant pressure gradient and therefore not truly applicable to separated flows. Thus the application of the instantaneous wall laws is not without doubt.

Breuer et al. (2007) and Kalitzin et al. (2008) provide a good summary of the recent developments of wall models for LES of turbulent flows. Breuer et al. (2007) have also developed a wall model for three-dimensional complex flows with large pressure gradients and demonstrated the efficacy of their model for the test case of periodic hill flow.

As we have already seen, a wall model provides expression for the flow quantities in the wall region rather than actually accurately computing them. The following expressions are used for the mean velocity in the wall laws (Benarafa et al. 2007)

$$u^+ = y^+ \quad \text{for } y^+ < 5 \tag{8.12}$$

$$u^+ = \int_0^{y^+} \frac{2d\alpha}{1 + \sqrt{1 + 4lm_+^2(\alpha)}} \quad \text{if } 5 < y^+ < 30 \tag{8.13}$$

$$\text{With } lm_+(\alpha) = k\alpha^+ \left(1 - \exp\left(-\frac{\alpha}{26}\right)\right) \text{ in the buffer layer} \tag{8.14}$$

$$u^+ = \frac{1}{k}\ln(y^+) + A \text{ if } y^+ > 30 \text{ (logarithmic layer)} \tag{8.15}$$

Here all variables refer to the instantaneous behaviour. The instantaneous velocity is normalized with the friction velocity and y_w denotes the distance normal to the wall.

To model the wall normal heat flux the Kader law (1981) is used for the instantaneous temperature

$$\frac{T - T_w}{T_\tau} = \Pr y^+ e^{-\Gamma} + [2.12 \ln[(1 + y^+)C] + \beta]e^{-1/T} \tag{8.16}$$

$$\text{With } \Gamma = \frac{10^{-2}(\Pr y^+)^4}{1 + 5\Pr^3 y^+} \quad \text{and} \quad C = \frac{1.5(2 - y/h)}{1 + 2(1 - y/h)^2} \tag{8.17}$$

The temperature is normalized by the friction temperature T_τ which is defined as

$$T_\tau = \frac{\alpha}{u_\tau}\left(\frac{\partial T}{\partial y}\right)_w = \frac{\phi_w}{\rho C_p u_\tau} \tag{8.18}$$

Here all symbols have the standard meanings.

8.7.2 Velocity and Temperature TBLE Wall Model

A major assumption and limitation of the thin boundary-layer approach is that the interaction between the inner and outer regions is weak. On the inner mesh the following thin boundary layer equations are solved for the velocity field (Benarafa et al. 2007).

$$\frac{\partial u_i}{\partial t} - \frac{\partial}{\partial y}\left\{ (v + v_t)\frac{\partial u_i}{\partial y} \right\} = -\frac{\partial P}{\partial x_i} - \frac{\partial (u_i u_j)}{\partial x_j} \tag{8.19}$$

The mean velocity in the wall-nearest cell obtained from the LES computations in the outer region provides an outer boundary condition for Eq. 8.19 in addition to the no slip condition at the wall. In other words, the momentum equation for thin boundary layer region is solved subject to the above mentioned two boundary conditions. The time averaged shear stress obtained from the Reynolds-averaged equations is directly used as the boundary condition for LES computations. Some minor modifications to these treatments have been proposed in the literature.

Since the Poisson equation is not solved, the normal velocity is obtained by integrating the mass conservation equation

$$v(y) = - \int_0^y \left\{ \frac{\partial u}{\partial x}(\alpha) + \frac{\partial w}{\partial z}(\alpha) \right\} d\alpha \tag{8.20}$$

The normal velocity is used to compute the convection terms in Eq. 8.19. The eddy viscosity in Eq. 8.19 is computed from the mixing length model using a van Driest damping function.

$$v_t = D_{vd}(\kappa y)^2 \left\{ \left(\frac{\partial u}{\partial y} \right)^2 + \left(\frac{\partial w}{\partial y} \right)^2 \right\}^{1/2} \tag{8.21}$$

The van Driest damping function in Eq. 8.19 is given as

$$D_{vd} = 1 - \exp\left[-\left(\frac{y^+}{25} \right)^3 \right] \tag{8.22}$$

The temperature field in the inner mesh is obtained by solving the following transport equation for thermal energy

$$\frac{\partial T}{\partial t} - \frac{\partial}{\partial y}\left\{ (\alpha + \alpha_t)\frac{\partial T}{\partial y} \right\} = Q - \frac{\partial (T u_j)}{\partial x_j} \tag{8.23}$$

Here Q denotes a source of thermal energy. The turbulent Prandtl number (Pr_t) is assumed to be approximately equal to one and turbulent thermal diffusivity is assumed to be equal to the eddy viscosity.

It can be shown that though overall this model is quite accurate but it requires some improvements to remove the problem of mean velocity mismatch at the interface of the inner and outer regions and spurious location of the fluctuation peak at the interface.

8.8 Initial, Boundary Conditions and Duration of Computations

As we have already seen a key issue in DNS and LES is the specification of the initial and boundary conditions and the duration up to which the computations should be performed. Usually the initial conditions for a particular flow configuration are taken from a similar computation for a similar geometry. However, in case such data is not available because a new flow configuration is being investigated, a random flow field superimposed on a mean flow field is used as the initial conditions and the computations are performed for few eddy turn over times. The flow field thus obtained is in turn used as the initial condition for the next (and perhaps final) round of computations. The data generated by performing DNS or LES based on such initial conditions is likely to be more realistic compared to the random initial field taken earlier.

Open boundary conditions pose the maximum difficulty in DNS and LES. Because of the elliptic nature of the governing equations the flow upstream depends on the downstream flow behaviour and therefore an accurate representation of such conditions at an open boundary becomes a challenging task. Unsteady convection is one of the commonly used open boundary conditions and is given as

$$\frac{\partial \phi}{\partial t} + u\frac{\partial \phi}{\partial n} = 0 \qquad (8.24)$$

Here ϕ denotes a flow variable. The outflow velocity U is selected in such a way so that the overall mass conservation is maintained. This condition essentially means the outflow mass flux is equal to incoming mass flux.

The boundary conditions on a solid wall are straightforward (the no slip). If a particular flow property does not vary in a specific direction, the variation of this flow property in that direction can be assumed to be zero.

Once the best possible initial and boundary conditions have been selected, the next question that arises is as to what is the time duration to which the computations should be performed, because the computational cost is directly proportional to the duration of the run. The computations should be performed for at least few eddy turn over times after which the time-averaging of the computed flow field can be performed. The eddy turn over time is essentially the integral time scale of the flow and is defined as the integral length scale divided by the mean square velocity fluctuations. Integral length scale of turbulence is approximately equal to the distance over which the fluctuating components of the velocity are correlated and this is a measure of typical eddy size.

Another issue that needs to be considered is as to what is the minimum size of the domain that should be considered so that all essential features are captured accurately. It should be at least equal to few times the integral length scale, so that large eddies are captured.

We will illustrate the above-mentioned issues by considering an example of turbulent flow past a cylinder of finite height mounted on a flat plate. This is a complex flow situation. Firstly the flow over the finite end affects the recirculation region behind the cylinder. The vortex shedding near the tip is also affected due to the finite height of the cylinder. Secondly the boundary-layer developing on the bottom plate creates horse shoe vortex around the cylinder base. We will see initial and boundary conditions and size of the computational domain that needs to be considered if we need to perform LES of this flow (Froehlich et al. 2003). The incoming flow considered by Froehlich et al. (2003) had a Reynolds number of 43,000 based on the diameter of the cylinder and incoming free stream velocity and the ratio of the height of the cylinder to its diameter was 2.5. The extent of the computational domain chosen by them in the streamwise direction was $20D$, in the spanwise direction $7D$ and in the wall normal direction $5D$, where D denotes the cylinder diameter. At the inlet plane which was located $7.5D$ upstream of the cylinder, a steady laminar flow was provided as the inflow condition. In other words, the flow underwent laminar to turbulent transition before it touches the cylinder. At the outflow plane, a convective flow condition was employed. In order to reduce the computational requirements they assumed slip at the bottom plate and showed that this simplification does not affect the flow near the free end of the cylinder. Froehlich et al. (2003) used the Smagorinsky subgrid scale model with the model constant $C_s = 0.1$. In a flow of this type it is important to accurately resolve the shear layer separating from the side of the cylinder. They used second order central difference scheme for the spatial discretization and three-step Runge–Kutta method for the temporal discretization and they obtained good agreement between their predictions and the corresponding experimental data of mean and turbulent flow fields.

8.9 Concluding Remarks

Flow field obtained from DNS comes closest to an actual raw turbulent data. It does not involve any assumption or simplification. A DNS data contains far more information than actually needed by engineers. However, it may contain very useful information. For example, we have already seen that the exact transport equation for rate of dissipation of turbulence kinetic energy contains several immeasurable correlations. A DNS data can be used to propose correlations to model different terms of such equations and thus to arrive at new turbulence models.

It is to be noted that there is no reason to believe that either DNS or LES performed on a given mesh and time step will be sufficiently accurate. Usually one starts the computations by employing a particular mesh that appears to be the best. Once the solution has been obtained over a sufficiently large real time, one assesses the accuracy of the results by checking the grid size in the wall region using the wall co-ordinates (which in turn depends on the friction velocity and which is part of the computed flow field). A good LES is the one in which the wall region is accurately resolved both spatially and temporarily. If after checking the grid spacing in the wall co-ordinates it turns out that a coarse mesh is used, the computations should be repeated using a fine mesh.

References

Balaras E, Benocci C, Piomelli U (1996) Two-layer approximate boundary conditions for large-eddy simulation. AIAA J 34:1111–1119

Bardina J, Ferziger JH, Reynolds WC (1980) Improved subgrid scale models for large eddy simulation. AIAA J 80:1357

Benarafa Y, Cioni O, Ducros F, Sagaut P (2007) Temperature wall modeling for large-eddy simulation in a heated turbulent plane channel flow. Int J Heat Mass Transfer 50:4360–4370

Breuer M, Knaizev B, Abel M (2007) Development of wall models for LES of separated flows using statistical evaluations. Comp Fluids 36:817–837

Froehlich J, Rodi W, Dewan A, Fontes JP (2003) Large eddy simulation of the flow around the free end of a circular cylinder, numerical flow simulation III. In: Hirschel EH (ed) Notes on numerical fluid mechanics and multidisciplinary design, vol 82. Springer, New York, pp 191–202

Germano M, Piomelli U, Moin P, Carbott WH (1991) A dynamic subgrid scale eddy viscosity model. Phys Fluids 3(7):1760–1765

Hanjalic K (2005) Will RANS survive LES? A view of perspectives. Trans ASME J Fluids Eng 127:831–839

Kader BA (1981) Temperature and concentration profiles in fully turbulent boundary layers. Int J Heat Mass Transfer 24(9):1541–1544

Kalitzin G, Medic G, Templeton JA (2008) Wall modeling for LES of high Reynolds number channel flows: what turbulence information is retained? Comp Fluids 37:809–815

Piomelli U, Balaras E (2002) Wall-layer models for large eddy simulations. Ann Rev Fluid Mech 34:349–374

Reynolds WC (1990) The potential and limitation of direct and direct- and large-eddy simulations. In: Lumley JL (ed) Whither turbulence: turbulence at the crossroads. Springer, Berlin, pp 313–342

Smagorinsky J (1963) General circulation experiments with the primitive equations I. The basic experiment. Mon Weather Rev 91:99–165

Spalart PR, Allmaras SR (1992) A one equation turbulence model for aerodynamic flows. AIAA paper 92-439, Reno, USA

Spalart PR, Jou W-H, Stretlets M, Allmaras SR (1997) Comments on the feasibility of LES for wings and on the hybrid RANS/LES approach. In: Advances in DNS/LES. Proceedings of first AFOSR international conference on DNS/LES

Werner H, Wengle H (1991) Large-eddy simulation of turbulent flow over and around a cube in a plate channel. In: Proceedings of 8th symposium turbulent shear flows, vol 19, no 4, pp 155–165

Chapter 9
Some Case Studies

Abstract This chapter presents some examples of turbulent flows encountered in industry. We see how they have been successfully modeled and computed using some of the techniques presented earlier in this book. Turbulent flow fields in these examples have been treated using RANS based models of different complexities as well as more advanced techniques such as DNS and LES. The examples cover a wide spectrum involving heat exchangers with turbulent convective heat transfer, stirred vessel with complex three-dimensional, turbulent and rotating flow, turbulent flow in tundish used in steel industry, ventilation in buildings, film cooling effectiveness. A reader may like to read the corresponding papers for further details of the cases presented.

9.1 Heat Exchangers

Heat exchangers are the equipment used for an efficient transfer of thermal energy from a hot fluid to a cold fluid, in most cases through an intermediate metallic wall and without any moving parts. Main types of heat exchangers can be grouped according to the flow layout: Shell-and-tube heat exchanger, plate heat exchanger, open-flow heat exchanger, and contact heat exchangers. In terms of the hydraulic diameter (D_h) of the flows inside a duct, heat exchangers may be grouped as: Conventional or non-compact heat exchangers, if $D_h > 5.0$ mm, compact heat exchangers, if 1.0 mm $< D_h < 5.0$ mm, meso heat exchangers if 0.2 mm $< D_h <$ 1.0 mm; and micro heat exchangers if $D_h < 0.2$ mm.

It is the goal of designers to enhance the heat transfer with the minimum penalty, i.e., correspondingly small increase in the pumping power. The flow in a typical heat exchanger is quite complex mainly due to the complex geometries involved. The complexity is enhanced due to the flow being turbulent. Figure 9.1 shows a typical pin fin heat sink used for the cooling of electronic appliances and

A. Dewan, *Tackling Turbulent Flows in Engineering*, 105
DOI: 10.1007/978-3-642-14767-8_9, © Springer-Verlag Berlin Heidelberg 2011

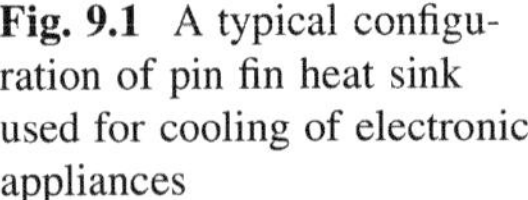
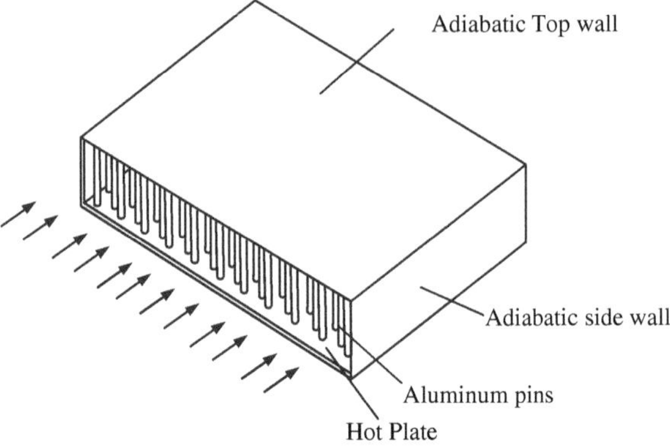

Fig. 9.1 A typical configuration of pin fin heat sink used for cooling of electronic appliances

Fig. 9.2 shows simplified model that may be used for the numerical simulations of such heat exchanger (Dewan et al. 2008).

In Fig. 9.2, the symmetry conditions on the two side walls have been imposed to reduce the size of the computational domain. This simplification is reasonable as the number of fins in the spanwise direction is quite large. Two solid walls are present in the computational domain (one on the top and one at the bottom) and therefore the specification of the boundary conditions on two walls for all flow quantities is straightforward: no slip for the flow field, temperature gradients in the direction normal to the walls being zero in case it is an adiabatic wall and turbulence zero at the wall. The bottom wall temperature is usually specified in case its temperature is available otherwise heat flux available is specified. Dewan et al. (2008) used the wall function approach to reduce the computational cost while treating turbulence in the vicinity of the solid wall. In addition to the heat exchanger some additional lengths at the inlet and exit of the computational domain need to be considered with corresponding suitable conditions. This is because at the actual inlet and outlet of the actual heat exchanger either realistic (experimentally measured) conditions should be used or if such conditions are not

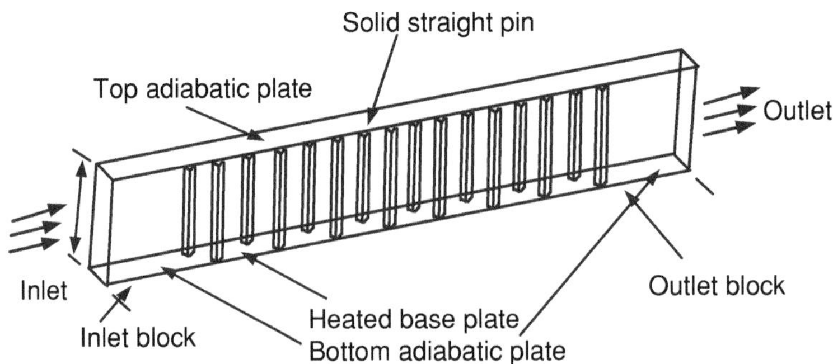

Fig. 9.2 Flow configuration of typical circular pin fin compact heat exchanger used in the cooling of electronic appliances

available, then some physically suitable conditions should be used. Therefore, Dewan et al. (2008) used uniform inflow condition at the inlet plane located upstream of the heat exchanger (Fig. 9.2). This specification allows the development of boundary-layers at the bottom plate before the flow hits the first pin fin. They located the outflow plane far downstream of the last pin fin and used the condition of the zero streamwise gradients of the flow properties at the outlet plane. Clearly this condition cannot be used if the outlet plane is located just downstream of the last pin fin.

Dewan et al. (2008) have shown that RANS based two equation models are capable to predict the overall behavior of such heat exchangers. In addition to three-dimensional turbulent convective heat transfer to air within the computational domain, heat is also transferred by conduction through the pin fin from the bottom heated plate and this phenomenon results in a temperature variation in a pin fins from the bottom towards their top. This variable fin temperature along its height in turn affects the convective heat transfer through air and necessitates simultaneous solutions of convective heat transfer equations through air and conduction heat transfer through solid pin fins (termed as the conjugate heat transfer).

9.2 Stirred Vessels

Stirred vessels are widely used in the chemical, mineral processing, wastewater treatment and several other industries. A large number of process applications involve mixing of single-phase fluid in mechanically stirred vessels. In a stirred vessel, a rotating impeller interacts with the fluid resulting in complex three-dimensional unsteady, rotating flow in the vessel and consequently into mixing of fluid. The mixing in such vessels is due to two phenomena: bulk fluid motion and turbulent diffusion. While the former leads to global mixing within the vessel, the latter leads to the local mixing. The flows encountered in the industry usually are at a very high Reynolds number and therefore, as already mentioned, due to enormous computational cost involved, DNS and LES cannot be used a design tool for such flows. Consequently RANS based turbulence models need to be used to investigate such flows and there is a need for choosing a suitable turbulence model which can represent the complex flow physics as accurately as possible. Another important issue is the treatment of the rotating motion of the impeller in the stationary vessel. Three approaches can be used to treat this rotating motion: (a) black box approach; (b) multiple reference frames (MRF); and (c) sliding grid. In the first approach, a small region around the impeller is treated as a black box. Experimentally measured profiles are imposed at boundaries of box surrounding impeller in the tank region to obtain the flow field in the vicinity of the impeller. However in this approach the flow behavior at the boundaries of a specific impeller needs to be obtained experimentally and therefore this approach cannot be applied for a new impeller for which the experimental data is not available. In the second approach (MRF) the governing equations are solved using a rotating frame in impeller region and stationary frame in tank region. An exchange of information at the interface between two regions takes place. This is the most widely used approach. In the sliding grid a time dependent moving and deforming grid is used to resolve unsteady flow structures. This approach is quite accurate, but computationally expensive and therefore cannot be used as a design tool.

The computed mixing time in a stirred vessel is obtained from the solution of the unsteady scalar transport equation after obtaining the turbulent flow field. A small amount of a tracer is injected somewhere in the vessel and the average time of its mixing at different points within the vessel is taken as the averaged mixing time. Dewan et al. (2006) have compared the performance of six turbulent models for the mixing in a vessel stirred by a grid disc impeller which has the radial characteristics and shown that the standard k–ε model provides the best agreement with the experimental data for a vessel stirred by a grid disc impeller.

9.3 Flow in a Tundish Used in Steel Making

Tundish is a reservoir located between the ladle and mould in a continuous casting unit. Tundish holds molten steel before it is poured in a mould and therefore plays an important role in removing any undesirable inclusion particles resulting from sources like de-oxidation and re-oxidation products, slag entrapment, exogenous inclusions and many chemical reactions. Inclusions are lighter than the molten steel and therefore these may be removed either by floating to the top slag or by sticking to the tundish walls. Use of flow modifiers promotes the flow in the tundish towards the top surface which help the inclusions float and get trapped by the slag layer at the top surface (Jha et al. 2008).

Jha et al. (2008) considered six strand billet caster tundish. They solved three-dimensional mean turbulent flow field. The presence of solid walls and the interaction of the flow with dams of different heights add to the flow complexity. Jha et al. (2008) used the standard k–ε model with the standard wall functions to treat the solid walls. They considered only half of the tundish with the symmetry conditions at a plane in the streamwise direction located in the middle of tundish. They assumed the top surface to be a free surface. They assumed that the inclusions touching side walls are reflected, reaching the top surface are trapped and those reaching the outlet move out of the tundish.

Particle trajectories need to be obtained once the mean flow field has been obtained. Jha et al. (2008) obtained the inclusion trajectories using a Lagrangian particle tracking method, which solves a transport equation for each inclusion as it travels through the previously calculated steady flow field. They also considered the chaotic effect of turbulent eddies by including a random component of velocity to the time averaged flow field already obtained from the numerical solution of the governing time averaged Navier–Stokes equations. The mean local inclusion velocity components needed to obtain the particle path, were obtained from the force balance involving the drag and buoyancy forces. Jha et al. (2008) injected the inclusions at different locations distributed homogeneously over the inlet plane. Each trajectory was calculated through the constant steel flow field until the inclusion either is trapped on the top surface or exits the tundish outlet. They showed that the inclusion removal tendency increases with the use of dams as compared to the bare tundish. It was also the dams, inclusion removal tendency increases with increase in dam heights.

9.4 Turbulent Plume

As we have already seen that a plume is a buoyancy-driven free shear flow (for example, smoke arising from a chimney). Like other free shear flows this is highly unstable and undergoes transition to turbulent flow even at a small value of Grashof number. We can compute the time-averaged turbulent round plume behavior by using the boundary-layer assumptions and also Boussinesq approximation of buoyancy variation. The latter means that the variation in density is negligible everywhere except in the body force term in the momentum equation.

We will see how we can compute this flow using the k–ε model of turbulence. The governing equations for the steady time-averaged flow under the boundary-layer assumptions are

Continuity

$$\frac{\partial \bar{u}}{\partial x} + \frac{1}{r}\frac{\partial (\bar{v}r)}{\partial r} = 0 \tag{9.1}$$

Axial momentum

$$u\frac{\partial \bar{u}}{\partial x} + v\frac{\partial \bar{u}}{\partial r} = \frac{1}{r}\frac{\partial}{\partial r}\left\{r\left(-\overline{u'v'}\right)\right\} + g\beta(\bar{T} - T_\infty) \tag{9.2}$$

Thermal energy

$$u\frac{\partial \bar{T}}{\partial x} + v\frac{\partial \bar{T}}{\partial r} = \frac{1}{r}\frac{\partial}{\partial r}\left\{r\left(-\overline{v't'}\right)\right\} \tag{9.3}$$

Here all symbols have the standard meanings, $\bar{T}$ denotes the local fluid temperature within the plume and T_∞ the ambient temperature. β denotes the coefficient of volumetric expansion and G_k the production of turbulence due to buoyancy.

Turbulence kinetic energy

$$\bar{u}\frac{\partial k}{\partial x} + \bar{v}\frac{\partial k}{\partial r} = \frac{1}{r}\frac{\partial}{\partial r}\left(r\frac{v_t}{\sigma_k}\frac{\partial k}{\partial r}\right) + P_k + G_k - \varepsilon \tag{9.4}$$

Rate of turbulence kinetic energy dissipation

$$\bar{u}\frac{\partial\varepsilon}{\partial x} + \bar{v}\frac{\partial\varepsilon}{\partial r} = \frac{\partial}{\partial r}\left(r\frac{v_t}{\sigma_\varepsilon}\frac{\partial\varepsilon}{\partial r}\right) + C_{\varepsilon1}\frac{\varepsilon}{k}(P_k + G_k) - C_{\varepsilon2}\frac{\varepsilon^2}{k} \tag{9.5}$$

Temperature fluctuations

$$\bar{u}\frac{\partial\overline{t'^2}}{\partial x} + \bar{v}\frac{\partial\overline{t'^2}}{\partial r} = \frac{1}{r}\frac{\partial}{\partial r}\left(rc_T\frac{k^2}{\varepsilon}\frac{\partial\overline{t'^2}}{\partial r}\right) + P_t - \varepsilon_t \tag{9.6}$$

Here $P_t = -\overline{v't'}\frac{\partial T}{\partial r}$ and $\varepsilon_t = \alpha\overline{\frac{\partial t'}{\partial x_j}\frac{\partial t'}{\partial x_j}}$ denote the production and dissipation of temperature fluctuations, respectively. The transport Eq. 9.6 for the temperature fluctuations is similar in nature to that for turbulence kinetic energy 9.4. Rather than using a separate transport equation for rate of dissipation of temperature fluctuations Dewan et al. (1997) related it to the dissipation of turbulence kinetic energy by using the following relation.

$$\varepsilon_t = C_{T1}\varepsilon\frac{\overline{t'^2}}{k} \tag{9.7}$$

It is assumed here that the time scales of velocity fluctuations and temperature fluctuations are approximately the same and their ratio $C_{T1} = 1.25$. In any buoyancy driven turbulent flow the buoyancy production of turbulence is also significant and should be considered in addition to the shear production. All the standard models have been designed for non-buoyant flows and therefore suitable modifications need to be incorporated to such models for the buoyancy driven flows. It may be noted that due to the boundary-layer assumptions the streamwise gradients of all flow properties are negligible compared to those in the normal direction. Further all molecular viscosity terms in all equations have been neglected compared to the corresponding eddy viscosity terms.

The computational domain chosen should be a rectangular box surrounding the plume (Fig. 9.3). We need to choose an optimum size of the computational domain in such a way that appropriate boundary conditions can be applied. The width of the domain should be at least about two to three times the expected velocity half width at the last downstream location. The length of the computational domain should be equal to the downstream location up to which one is interested. The bottom boundary should be located at the plume discharge. The time-averaged flow can be assumed to be symmetric about the centerline and therefore we need to compute only one half of the plume. The symmetry condition means that the gradients of all variables normal to the axis are zero and from continuity it can be shown that the normal velocity is also zero at the centerline. Here the buoyancy arises due to a variation in density of air due to the heating. We can assume the temperature (T) at all boundaries (except the discharge region at the bottom boundary) to be equal to the ambient temperature (T_∞). At the right boundary the working medium can be assumed to be stationary and therefore no turbulence and

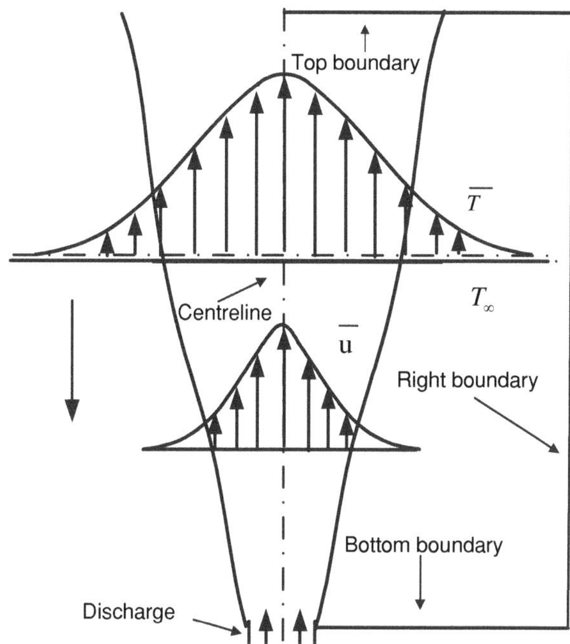

Fig. 9.3 Turbulent round plume

temperature of the medium is equal to the ambient temperature. Since the boundary-layer equations are solved which are parabolic in nature and therefore involve the marching solution, conditions at the top boundary are not required for the solution within the computational domain. Rather the flow properties at this boundary are a part of the numerical solution. At the discharge portion of the bottom boundary, the air temperature is equal to the heat source temperature, the flow velocity is equal to the discharge velocity with a small amount of turbulence. At the remaining part of the bottom boundary all flow variables are zero and air temperature is equal to the ambient value (T_∞).

9.5 LES and DES of Particle Deposition in a Human Throat

Medicines are inhaled by patients of asthma and obstructive pulmonary disease. The aerosol particles inhaled by a patient need to pass through complex geometry of mouth and throat before reaching the destination, i.e., lungs. Jayaraju et al. (2008) performed a computational study of deposition of different particles of diameters 2, 4, 6, 8 and 10 μm in a human throat model under the standard breathing condition of 30 l/min using RANS equations and low Reynolds number variation of shear stress transport k-ω model without near wall corrections. They used FLUENT 6.3 for computations and also considered detached eddy simulation based on Spalart and Allmaras model with the empirical constant $= 0.65$ and large eddy simulation. They considered two models for subgrid scales in LES, namely, Smagorinsky-Lilly model and wall adapting local eddy viscosity model in which the computed value of the eddy viscosity depends on the flow parameters in the presence of the wall. They compared their predictions with their experimental data using PIV. They showed that overall the LES is superior to the two other approaches. However, the predictions by DES were somewhat poorer than those of LES mainly because it involves the solution of additional eddy viscosity equation and the associated physical uncertainty with this approach. It is difficult to state how accurately they have resolved the turbulence especially in the wall regions. However, their computations seem to be fairly accurate from engineering view point.

They considered the Lagrangian equations governing the particle motion with the assumption of large particle and air density ratio, negligible particle rotation and negligible inter-particle collision. They assumed the drag force to be the dominant force. For particle tracking they used an eddy interaction model for RANS computations. In LES and DES particle phase iteration follows the fluid phase iteration and particles are tracked in real time. They also considered a simplified form of the LES for particle tracking termed as the frozen LES where particles are injected and tracked after a small interval, i.e., initially there is no non-linear interaction between fluid and particle. For RANS computations they meshed the geometry with 0.8 millions hexahedral cells with the location of first cell at y^+ approximately equal to 1 and for LES they used 1.9×10^6 hexahedral cells with the first cell located at y^+ approximately equal to 0.2 whereas for DES they used 1.9×10^6 hexahedral cells with the first cell located at y^+ approximately equal to 1. They concluded that LES approach though quite accurate may not be realistic for simulating transient inhalation/exhalation cycle. They observed a large difference between the values of the particle deposition computed by RANS and LES.

9.6 Unsteady Cross Ventilation in Buildings

Fluctuations in incoming wind flow can result in a significant enhancement in natural ventilation within a building mainly because it causes unsteadiness in approaching shear layer. Hu et al. (2008) performed one such study by considering two wind directions, one parallel to the direction of opening and other normal to an opening in a simple model of a building. They used the simplest subgrid scale model of Smagorinsky (1963) with Smagorinsky's constant $= 0.12$ and employed Werner and Wengle (1991) two-layer model to resolve the wall region. A key issue in the prediction of flow past tall buildings is the generation of accurate inflow conditions and usually this is done by considering a driver section (an additional computational domain) ahead of the domain containing the building. The flow behavior at the outlet of this driver domain is provided as the inflow condition to main computational domain. Hu et al. (2008) also considered a driver domain to generate fluctuating wind flow conditions in addition to the regular computational domain. They used the second order central difference scheme for the convection terms of the momentum equation. They observed a good agreement between their measurements of mean velocity, turbulence kinetic energy of the approaching wind and pressure coefficient within and outside the building and their LES computations. They suggested that due to inherent unsteady nature a deterministic approach may not be adequate for such studies. They showed that the fluctuations result in a significant enhancement in ventilation.

As we have seen from the example of Hu et al. (2008) that LES is appropriate for investigating the problems in wind engineering. The treatments based on RANS equations fail for such flows dominated by unsteady separation with vortex shedding (Tominaga et al. 2008 and Tamura 2008). Tominaga et al. (2008) also considered three different versions of the k–ε model modified to account for the effects encountered in flow past bluff bodies. The computations showed no vortex shedding even when unsteady computations using the abovementioned RANS models were performed thus demonstrating the limitations of RANS equations. Tamura (2008) presents an interesting compilation of following six cases dealing with wind engineering to show that LES is an effective tool: (a) wind flow over a hill with sinusoidal curve; (b) ground surface modeling for vegetation; (c) safety around a heliport located on a cliff top; (d) effect of wind in cities; (e) dispersion of hazardous gases in urban areas; and (f) meso scale meteorological model.

9.7 Effect of Turbulent Prandtl Number on Film Cooling

Industrial gas turbines are operated continuously over long periods of time. As a result components working at high temperatures (for example, such as turbine rotors and vanes) experience significant stresses. Therefore, gas turbine components need to be cooled and film cooling is used to maintain component temperatures within permissible limits. Liu et al. (2008) have used a two-layer model by employing the realizable k–ε model in the outer fully turbulent region and an isotropic one equation model in the near wall region which includes the viscous sublayer, buffer region and logarithmic region. The computational domain considered by them consisted of a cooling hole diameter of size $D = 2.35$ cm with a length-to-diameter ratio of 30 to ensure that the flow is fully developed and its velocity profile is close to that obtained in the measurements. The length of the channel upstream to the hole exit center was 22 D. They employed the time averaged governing equations for three-dimensional, steady flow and considered a variable turbulent Prandtl number and compared their predictions with the experimental data. They showed that a variable turbulent Prandtl number has a significant influence and a reduction in its value increases the film cooling effectiveness. They concluded that the use of a variable turbulent Prandtl number to some extent overcomes the limitation of a two equation model which arises due to the use of an isotropic eddy viscosity.

9.8 Flow Over Rough Walls with Suction

A common method adopted in the literature to treat turbulent flow in the presence of a solid wall is to use the wall functions approach. However, as we have already seen that the standard logarithmic law cannot be directly applied to rough walls and in fact this law needs to be modified. Gregoire et al. (2003) considered a two-dimensional wall roughness with suction using a two layer zonal model with RNG k–ε model used in the outer fully turbulent region. They showed that the logarithmic law can be used, but origin at the crest of the actual wall should be taken and the distance from this origin should be considered. They used a replacement of the viscous length scale by roughness height in modeling of the von Karman constant in the existing logarithmic laws for rough impermeable walls. They showed that the mean velocity profiles obtained by these two-dimensional computations were in good agreement with the semi-logarithmic law for impermeable rough walls.

9.9 Separated Convection Due to Backward Facing Step

The flow separation and reattachment almost always determines the key structure of the turbulent flow field and significantly influences the mechanism of heat transfer. This knowledge is needed for simulating and optimizing the performance of heat exchanging devices where such situations are encountered. Chen et al. (2006) studied the effects of step height on two-dimensional, turbulent, separated flow adjacent to a backward-facing step using the low-Reynolds number $k - \varepsilon - \overline{t'^2} - \varepsilon_t$ model proposed by Abe et al. (1994, 1995). They considered a variable value of the turbulent Prandtl number and showed that the overall behavior was captured well by the turbulence model. The peak values of the transverse velocity component become smaller as the step height increases. The maximum temperature increases as the step height increases. They showed that the bulk temperature increases more rapidly as the step height increases.

9.10 Concluding Remarks

In the preceding chapters we have presented salient features of different turbulence models. In this chapter, we have seen different situations, where modeling of turbulence and its simulation is the key issue. An attempt has been made to present a wide spectrum of turbulence models and simulations, i.e., different types of two equation models, applications involving heat transfer, and large eddy simulations. The models presented in earlier chapters have been exemplified by considering several examples in the present chapter. In the next chapter we will present a summary of important topics covered in different chapters.

References

Abe K, Kondoh T, Nagano Y (1994) A new turbulence model for predicting fluid flow and heat transfer in separating and reattaching flows—I. Flow field calculations, Int J Heat Mass Transf 37 (1):139–151

Abe K, Kondoh T, Nagano Y (1995) A new turbulence model for predicting fluid flow and heat transfer in separating and reattaching flows—II. Thermal field calculations. Int J Heat Mass Transf 38(8):1467–1481

Chen YT, Nie JH, Armaly BF, Hsieh HT (2006) Turbulent separated convection flow adjacent to backward-facing. Int J Heat Mass Transf 49:3670–3680

Dewan A, Arakeri JH, Srinivasan J (1997) A new turbulence model for the axisymmetric plume. Appl Math Model 21:709–719

Dewan A, Buwa VV, Durst F (2006) Performance optimizations of a grid disc impeller for mixing of single-phase flows in a stirred vessel. Trans IChemE Part A Chem Eng Res Des 84(A8):691–702

Dewan A, Dayanand L, Patro P (2008) Mathematical modelling and computation of three dimensional, turbulent, convective heat transfer in a heat exchanger with circular pin fins. In: Virtanen EN (ed) Applied mathematical modeling. Nova Science Publishers, Inc., USA, pp 273–296

Gregoire G, Marinet FM, Amand FJS (2003) Modelling of turbulent flow over a rough wall with or without suction. ASME J Fluids Eng 125:636–642

Hu CH, Obha M, Yoshie R (2008) CFD modeling of unsteady cross ventilation flow using LES. J Wind Eng Ind Aerodyn 96:1692–1706

Jayaraju ST, Brouns M, Lacor C, Belkassem B, Verbanck S (2008) Large eddy and detached eddy simulations of fluid flow particle deposition in a human mouth throat. Aerosol Sci 39:862–875

Jha PK, Rao PS, Dewan A (2008) Effect of height and position of dams on inclusion removal in a six strand tundish. ISIJ Int 48(2):154–160

Liu CL, Zhu HR, Bai JT (2008) Effect of turbulent Prandtl number on the computation of film-cooling effectiveness, Int. J. Heat Mass Transf 51:6208–6218

Smagorinsky J (1963) General circulation experiments with the primitive equations I., the basic experiment, Mon. Weather Rev 91:99–165

Tamura T (2008) Towards practical use of LES in wind engineering. J Wind Eng Ind Aerodyn 96:1451–1471

Tominaga Y, Mochida A, Murakami S, Sawaki S (2008) Comparison of various revised k–ε models and LES applied to flow around a high rise building model with 1:1:2 shape placed within the surface boundary layer. J Wind Eng Ind Aerodyn 96:389–411

Werner H, Wengle H (1991) Large-eddy simulation of turbulent flow over and around a cube in a plate channel. In: Proceedings of 8th Symposium. Turbulent Shear Flow, Technical University, Munich

Chapter 10
Conclusions and Recommendations

Abstract In the preceding chapters we have presented characteristics of turbulent flows in details. We have also seen a wide variety of methods that can be adopted to model turbulent flows, using the Reynolds-averaged Navier–Stokes equations, from the simplest zero-equation model to the most complex Reynolds stress and turbulent heat flux transport models. We have also seen the direct treatment of three-dimensional, instantaneous turbulent flow using the large eddy simulation and direct numerical simulation. We have shown that LES has a good potential as a design tool in engineering applications. We have also seen case studies of some representative turbulent flows covering a wide spectrum. In the present chapter we will summarize the important observations from the previous chapters. We will also present some guidelines for adopting a particular approach for turbulent flow modeling and simulation.

10.1 Tackling Turbulence

Turbulence is a complex phenomenon and therefore difficult to model mathematically. Turbulence is characterized by unsteadiness, three-dimensional behaviour, wide range of length and time scales, dissipation, fluctuation in flow properties, etc. The presence of walls, buoyancy, heat transfer and other effects add to the complexity in modelling of turbulence. A first step in modeling turbulent flows should be to gather as much information as possible about the basic characteristics, for example, values of the governing parameters, whether the time averaged flow is steady or unsteady, two-dimensional or three-dimensional, typical flow dimensions, etc. We must not forget that a raw turbulent flow is always three-dimensional and unsteady. One must also evaluate computational resources available, for example, processor and available disk space, etc., before planning a computational investigation of a particular turbulent flow. Subsequently one should select an appropriate method of the simulation and turbulence model to be

A. Dewan, *Tackling Turbulent Flows in Engineering*,
DOI: 10.1007/978-3-642-14767-8_10, © Springer-Verlag Berlin Heidelberg 2011

adopted. For example, if the flow is likely to be dominated by large scale unsteadiness, then RANS based turbulence models are not likely to be suitable. LES or DNS is likely to be suitable for such flows but these techniques are computationally expensive.

If our goal is to perform a computational study then our objectives should be clear. The first issue that needs to be considered is whether we are primarily interested in time averaged quantities of engineering interest, for example, in overall heat transfer, friction factor and pumping power, etc., or instantaneous flow properties. As we have already seen if the interest is in the former then the RANS equations with a suitable two equation model may be adequate. However, if we are primarily interested in the instantaneous flow field, then we will need to perform either DNS or LES. We have already seen a major limitation of DNS, i.e., a very large computational effort, and therefore it is clear that for all practical purposes LES should be performed for such flows (and not DNS).

We need to choose an appropriate shape and size of the computational domain where suitable boundary conditions can be applied. The number of boundary conditions required depends on the nature and number of flow variables (and consequently the governing equations) that are to be obtained. The number and nature of the governing equations depend on the flow variables of interest. It also depends on the nature of flow. Our goal should be to simplify the mathematical model as much as possible, but without compromising the physics of the flow. It is always a good idea to try to explore if either the symmetry boundary condition or the periodic boundary condition can be applied because these reduce the size of the computational domain and thus the computational effort. We must always bear in mind that a raw three-dimensional, unsteady, turbulent flow has no symmetry in any flow properties with respect to any plane or line. However, a time-mean flow may be symmetric and/or periodic. For example, we can assume time mean turbulent jet flow to be symmetric about the centerline. Likewise we can assume a time mean flow in a vessel stirred by a grid disc impeller to be symmetric about two vertical planes 90° apart and containing a baffle.

The presence of walls poses additional challenges and it is not clear which treatment (either wall functions approach or a low-Reynolds number versions) are to be used while modeling the flow using the RANS equations. As we have already seen, both these approaches have their shares of advantages and disadvantages. In addition, a major difference in the computational requirements between the two approaches is a factor that needs considerable thought before choosing a particular approach. In general, the wall function approach requires less computational effort compared to that for low-Re models.

The last but not the least important issue that needs to be resolved is to choose an appropriate turbulence model and as we have seen that, unfortunately, no universal turbulence model exists in the literature. In conclusion we can say that it is likely that no better RANS based turbulence model than the existing ones is likely to be developed in the foreseeable future. Further RANS based models are likely to be used for engineering flow computations. However, LES has a good potential to be used for the investigation of practical fluid flow problems.

10.2 CFD Issues

After having seen various issues related to mathematical modeling and simulation of turbulent flows, it is also important to look at some CFD issues as well. The first and foremost is the selection of a suitable mesh and the accuracy of the numerical discretization scheme employed. As we have seen in Sect. 2.6 in Chap. 2, a turbulent flow has sharp gradients close to a solid wall and therefore grid needs to be sufficiently fine close to a wall. Secondly, the space and time discretization schemes should be of sufficiently good accuracy. A reader may refer to related books on CFD (for example, Ferziger and Peric 2002; Patankar 1980; Tennehill et al. 1997; Veersteeg and Malalasekara 2007) for details of different numerical schemes and their accuracies.

For example, in DNS and LES one is interested in accurate transient results and therefore schemes that are appropriate for steady flows cannot be adopted. Further the time step adopted should be stable. In most cases, explicit time advance methods can be used because small time steps are used in LES and DES. However, near walls, due to extremely fine structures, an extremely small step size needs to be used and therefore the use of explicit time advance methods may result in numerical instabilities and therefore it may be a good idea to use implicit methods. Fourth order Runge–Kutta method is widely used for the time marching in DNS and LES. The spatial discretization should also be accurate enough. Grid independence and code validation are also essential before one's computations can be believed to be accurate.

References

Ferziger JH, Peric M (2002) Computational methods for fluid dynamics. Springer, USA
Patankar SV (1980) Numerical heat transfer and fluid flow. Hemisphere Publishing Corporation, USA
Tennehill JC, Anderson DA, Pletcher RH (1997) Computational fluid mechanics and heat transfer, vol 2e. Taylor and Francis, USA
Veersteeg HK, Malalasekara W (2007) An introduction to computational fluid dynamics. The finite volume method, vol 2e. Prentice Hall, England

Index

If you have any concerns about our products,
you can contact us on
ProductSafety@springernature.com

In case Publisher is established outside the EU,
the EU authorized representative is:
Springer Nature Customer Service Center GmbH
Europaplatz 3, 69115 Heidelberg, Germany

Printed by Libri Plureos GmbH
in Hamburg, Germany